An Intuitive Exploration of Artificial Intelligence

Simant Dube

An Intuitive Exploration of Artificial Intelligence

Theory and Applications of Deep Learning

 Springer

Simant Dube
Technology and Innovation Office
Varian Medical Systems
A Siemens Healthineers Company
Palo Alto, CA, USA

ISBN 978-3-030-68626-0   ISBN 978-3-030-68624-6   (eBook)
https://doi.org/10.1007/978-3-030-68624-6

This Springer imprint is published by the registered company Springer Nature Switzerland AG
The registered company address is: Gewerbestrasse 11, 6330 Cham, Switzerland

"Learn how to see. Realize that everything connects to everything else."

Leonardo da Vinci

To my parents, for a warm home with lots of books,
to Mini, my life partner, for your unwavering encouragement and support, sine quo non,
to Yuvika and Saatvik, for the pitter-patter sound of your little feet every morning of your childhood,
to Sandra, for all the thought-provoking discussions about physics and mathematics with you,
to Saundarya, for all the chess, carrom, and badminton chill-out time with you,
to all lovers of science and mathematics endowed with natural intelligence, which outperforms AI at the present,
and to all the amazing AI machines in the future who will exceed my abilities in all respects despite their artificial intelligence.

Preface

> "I visualize a time when we will be to robots what dogs are to humans, and I'm rooting for the machines."
>
> Claude Shannon

> "Artificial intelligence would be the ultimate version of Google. The ultimate search engine that would understand everything on the web. It would understand exactly what you wanted, and it would give you the right thing. We're nowhere near doing that now. However, we can get incrementally closer to that, and that is basically what we work on."
>
> Larry Page

In June 2014, we drove to the Computer History Museum in Mountain View, California, not too far from where I was living. A spectacular surprise was in store for us. As soon as we entered the building, our eyes were riveted on an 11 ft. (3.4 m) long and 7 ft. (2.1 m) high sculpture of engineering weighing five tonnes and having 8000 parts. One hundred and sixty-five years after it was originally conceived and designed, the Difference Engine 2 of Charles Babbage was operational in front of us. I wonder if the Cambridge mathematician from Victorian England could foresee that one day his work would be on display in the heart of Silicon Valley. We listened with rapt attention to the docent talk and demonstration of this marvelous machine.

"I wish to God these calculations had been executed by steam!" an anguished Charles Babbage exclaims as he finds error after error in astronomy tables calculated by human hand. It is London and the summer of 1821, and this statement is a remarkably prescient wish for an era of automation by machines that would carry out intelligent tasks without any errors. After 200 years, Larry Page's vision of the ultimate version of Google is not too different. Instead of the goal of automating tedious calculations, Google strives to build machines that will automatically understand everything on the World Wide Web.

What has changed in the intervening two centuries is the ubiquity of computing that has opened the doors to an unstoppable, colossal flood of data, also known as big data. Within the big data are numbers, observations, measurements, samples, metrics, pixels, voxels, 3-D range data, waveforms, words, symbols, fields, and records—which can all be processed and munged to expose structures and relationships. We want to use the big data to predict, classify, rank, decide, recommend, match, search, protect, analyze, visualize, understand, and reveal.

Of course, we are the creators of data, and we understand the underlying processes. We can bring our expertise to the fore and employ it to build machine learning models. This is the basis of classical ML, the parent of AI. Classical ML demands human labor during a process called feature engineering. It can work remarkably well for some applications, but it does have limits.

AI seeks to learn from the raw data. It chooses a large, deep neural network with tens or hundreds of millions of internal learnable parameters that are tweaked during training using calculus to minimize a suitably chosen loss function on a large annotated training dataset consisting of tens or hundreds of millions of examples. Once tweaking has been completed, the network can be deployed in production and used in the field to carry out the magical "intelligent" inference. During inference, AI models execute billions of multiplications, additions, and comparisons per second. If we were to create a mechanical slow-motion replica of such an AI model and witness a docent demonstration in a futuristic museum, it would truly be mind-boggling to see the countless moving parts of the endlessly sprawling machine as it crunched out a final answer.

This book is an endeavor to describe the science behind the making of such an impressive machine. One may ask—where is intelligence in the middle of billions of clinking, clanking parts? For human intelligence, we have first-hand subjective experience and we demand no further proof. For AI, the question is a difficult one. Let us take the first step of understanding how AI works.

Berkeley, CA, USA   Simant Dube
January 2021

Contents

Acronyms

L^p	Lebesgue Space.
AGI	Artificial General Intelligence.
AI	Artificial Intelligence.
ALS	Alternating Least Squares.
ASIC	Application-Specific Integrated Circuit.
ASR	Automatic Speech Recognition.
AUC	Area Under Curve.
BERT	Bidirectional Encoder Representation from Transformer.
Capsnet	Capsule Neural Network.
CART	Classification and Regression Trees.
CNN	Convolutional Neural Network.
CPU	Central Processing Unit.
CTC	Connectionist Temporal Classification.
DBN	Deep Belief Network.
DBSCAN	Density-Based Spatial Clustering of Applications with Noise.
DNC	Differentiable Neural Computer.
DNN	Deep Neural Network.
DQN	Deep Q-Network.
DRL	Deep Reinforcement Learning.
E2E	End-to-End.
EM	Expectation Maximization.
FCN	Fully Convolutional Network.
FDA	Fisher Discriminant Analysis.
FFN	Feed-Forward Network.
FPGA	Field-Programmable Gate Array.
FPR	False Positive Rate.
GAN	Generative Adversarial Network.
GBM	Gradient Boosted Machine.
GPR	Gaussian Process Regression.
GPT	Generative Pretrained Transformer.
GPU	Graphics Processing Unit.

GRU	Gated Recurrent Unit.
HMM	Hidden Markov Model.
HOG	Histogram of Oriented Gradients.
IOU	Intersection Over Union.
JS	Jensen-Shannon.
k-NN	k-Nearest Neighbor.
KL	Kullback-Leibler.
LAD	Least Absolute Deviation.
LDA	Linear Discriminant Analysis.
LSTM	Long Short-Term Memory.
MAP	Maximum a Posteriori.
mAP	Mean Average Precision.
MART	Multiple Additive Regression Trees.
MCMC	Markov Chain Monte Carlo.
MCTS	Monte Carlo Tree Search.
MDP	Markov Decision Process.
MDS	Multidimensional Scaling.
MFCC	Mel Frequency Cepstral Coefficients.
ML	Machine Learning.
MLE	Maximum Likelihood Estimation.
MLP	Multi-Layer Perceptron.
MOE	Mixture of Experts.
MSE	Mean Squared Error.
NAS	Neural Architecture Search.
NCF	Neural Collaborative Filtering.
NDCG	Normalized Discounted Cumulative Gain.
NER	Named-Entity Recognition.
NLP	Natural Language Processing.
NLU	Natural Language Understanding.
NMS	Non-Maximum Suppression.
NMT	Neural Machine Translation.
NTM	Neural Turing Machine.
OCR	Optical Character Recognition.
OLS	Ordinary Least Squares.
PCA	Principal Component Analysis.
PGM	Probabilistic Graphical Model.
PLS	Partial Least Squares.
PN	Policy Network.
PR	Precision Recall.
Q	Quality (in reinforcement learning).
QA	Question And Answer.
QDA	Quadratic Discriminant Analysis.
R-CNN	Region-based CNN.
RANSAC	Random Sample Consensus.
RBM	Restricted Boltzmann Machine.

ReLU	Rectified Linear Unit.
RL	Reinforcement Learning.
RNN	Recurrent Neural Network.
ROC	Receiver Operating Characteristic.
SARSA	State, Action, Reward, State, Action.
Seq2Seq	Sequence-to-Sequence.
SGD	Stochastic Gradient Descent.
SIFT	Scale-Invariant Feature Transform.
SOFM	Self-Organizing Feature Maps.
SRL	Statistical Relational Learning.
SSD	Single-Shot Detector.
SVD	Singular-Value Decomposition.
SVM	Support Vector Machine.
t-SNE	t-Stochastic Neighbor Embedding.
TD	Temporal Difference.
TPR	True Positive Rate.
TPU	Tensor Processing Unit.
VAE	Variational Autoencoder.
VQ	Vector Quantization.
XGBoost	Extreme Gradient Boosting.
YOLO	You Only Look Once.

Author's Note

This book is about developing conceptual understanding of Artificial Intelligence (AI), a rapidly moving field of growing complexity and of immense transformative potential. Deep learning has emerged as a key enabling component of AI. Significant attention is being paid to it by the media, governments, businesses, and the public. It can safely be said that AI will continue to remain an exciting frontier for humanity to seek and explore for centuries to come. Besides the motivation of using AI for its myriad practical applications, we humans like to contemplate and comprehend. This book seeks an "understanding" of AI and Machine Learning (ML) in the truest sense of the word. It is an earnest endeavor to unravel what is happening at the algorithmic level, grasp how applications are being built, and show the long adventurous road in the future. It is my ardent wish to answer some fundamental questions. Can we really create intelligent machines? Can an algorithm serve as a model of learning and intelligence? Will there always be some differences between human intelligence and AI? In order to answer these questions, the first step is to understand the state of the art in AI.

The book rests on a vast knowledge store in the form of articles, books, interviews, news stories, lectures, and software, which are the result of diligent arduous work by many scientists, mathematicians, engineers, educators, students, authors, and journalists. I acknowledge their contributions, as the book is really a distillation of their work. While writing the book, I felt like a small child standing next to all the great minds and watching the beckoning waves of future breakthroughs while they break against the present shore.

The primary goal of the book is to serve as a textbook for AI and deep learning at college and university level for students who have basic familiarity with calculus, linear algebra, and probability and statistics. When being used as a textbook, the book follows inverting-the-classroom philosophy, which states that learning is an active process of discovery, research, and contemplation. The student is encouraged to look up references in the bibliography to do exercises and to dive deeper into topics. An ideal course in AI will use it as a textbook and complement it with course projects and paper-reading sessions. The book emphasizes deep learning because it has revolutionized the field of AI. When combined with other techniques such as reinforcement learning, it can lead to amazing results.

At the same time, the book is intended for anyone who wishes to learn about AI. Learning is lifelong, and I have met many people who want to keep themselves continually challenged by learning new fields. Mathematical notation is simplified, and various illustrations are used to convey the intuition. Because the target audience is anyone who wants to learn the fundamentals of AI, I have chosen geometrical insight and intuitive exposition as the guiding principles. Therefore, the writing style is often informal and based on analogies. The language of mathematics is employed to convey concepts with greater clarity when it is needed. The book emphasizes both the theoretical and applied sides of AI and, therefore, will be able to meet different learning needs.

This book project is a by-product of my own professional journey in computer science as the field grew, morphed, and changed over time. It is a personal journey of discovery and learning and of being gently nudged by my life partner, Mini, to share with the wider world the intuition behind this very fine application of mathematics and engineering. I hope you will enjoy the book.

Berkeley, CA, USA   Simant Dube
January 2021

Part I
Foundations

Nothing in life is to be feared, it is
only to be understood. Now is the time
to understand more, so that we may
fear less.

Marie Curie

Chapter 1
AI Sculpture

In every block of marble I see a statue
as plain as though it stood before me,
shaped and perfect in attitude and
action. I have only to hew away the
rough walls that imprison the lovely
apparition to reveal it to the other eyes
as mine see it.

Michelangelo

I choose a block of marble and chop
off whatever I don't need

Auguste Rodin

Michelangelo could see an angel trapped in a block of marble ready to be sculpted out of it. Rodin knew what to chop off to turn a marble block into a deeply engrossed thinker (Fig. 1.1).

The two artists worked with 3-D objects. In AI, when we want to classify and categorize, we work with multidimensional mathematical objects of great complexity. Unlike the two sculptors, we fail to visualize these objects and we resort to computer experimentation, a trial-and-error process prone to pitfalls and mistakes. We don't have access to the geometry of these objects. The only help we have comes in the form of a set of points belonging to the desired sculpture. It is a high-dimensional space, so despite being given this set of points, we still can't infer the geometry well. We do our best to carve out the sculpture so that the provided points are contained inside our chiseled object, hoping that it is a close approximation of the true sculpture.

Fig. 1.1 The Thinker by Auguste Rodin (model 1880, cast 1904). AI carves out sculptures in high-dimensional spaces for challenging problems, where the number of dimensions can be in hundreds of thousands. Every neuron performs a chiseling cut. It is a very fine process, which has led to the AI revolution. In this book, we will develop the intuition behind this carving process. The more intricate the carving is, the higher the expressive power of the AI model is. We will see that AI learns to chisel based on data and gradually improves itself with training (Courtesy the National Gallery of Art, Washington DC)

1.1   Manifolds in High Dimensions

The mathematical objects are manifolds or, more precisely, topological manifolds. Intuitively, manifolds are generalizations of curves and surfaces with the locally Euclidean property to higher dimensions. A curve is a 1-D manifold if for every point on the curve there exists a neighborhood around the point on the curve which is homeomorphic (of the same form) to an open interval (an interval without its endpoints) on the real line. Earth's surface is a 2-D manifold, because it is locally flat in a neighborhood around each point. The neighborhood is homeomorphic to the 2-D plane. The local neighborhood can be mapped using two axes, each point specified by two coordinates. Earth itself is a 3-D manifold with a boundary, where its surface is the boundary.

> *AI carves out manifolds for classification.*

Cats versus not-cats is an introductory classification task to teach AI to students. You are given many images of cats and of things and animals other than cats. Cats constitute the positive class and not-cats constitute the negative class. All possible

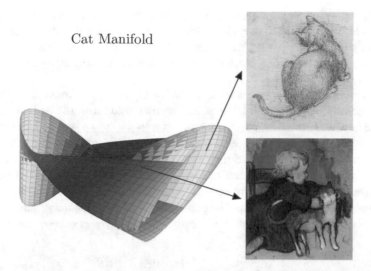

Cat Manifold

Fig. 1.2 Color (RGB) cat images of width W and height H form a manifold in a high-dimensional space. The number of dimensions N in the space is three times the total number of pixels, that is, $N = 3WH$. The goal of AI is to carve out this manifold given only a set of training points belonging to the manifold and to its complement. Note that a cat image is a single point in this manifold. The bottom image also belongs to the "child" manifold. (Top right: Leonardo da Vinci (c. 1513); bottom right, Paul Gauguin (c. 1890). Public Domain)

images of cats will form a manifold in a very high-dimensional space with each point corresponding to an image of some cat. The complement of the manifold will be the negative class. A point in the complement corresponds to some negative image.

In order to confirm that images of cats form a high-dimensional manifold, construct this space by using the most fundamental parts of an image, which are its pixel color or grayness values, as the axes. Let each image have exactly the same width W and height H. If we restrict our consideration to grayscale images, then there is one axis for each pixel and in total there are $N = W \times H$ dimensions. For color (RGB) images, there will be three times as many. Each pixel has three primary colors, namely, red, green, and blue. See Fig. 1.2. For 1 megapixel color images, we are talking about three million dimensions. We can easily navigate in space-time, which is a 4-D space (three for space and one for time). It is a totally different matter when it is three-million-dimensional space. We have a hard time visualizing just five dimensions. Our common intuitions about space and geometry break down rapidly in such spaces.

We call this space raw pixel space, where the qualification *raw* is added to emphasize that it is unprocessed, basic, and low level. A cat is an N-dimensional point $x \in \mathbb{R}^N$ in it. Given a cat's image, perturb it very slightly. It will be still a cat. There is a little room to jiggle the point x and preserve the property of being a cat. Mathematically speaking and in topological terms, it is an open neighborhood centered at the point, such that all neighboring points are still cats. The entire set of all cat points is an open set embedded in N-dimensional Euclidean space. It is a manifold. It is an angel in a block of marble ready to be sculpted by the engineers of AI.

What would be like to take a walk inside the cat manifold from one point x_A to another point x_B along a path lying inside the manifold? It would be like watching a movie in which the cat x_A gradually morphs into the cat x_B, with all intermediate images being of cats. If we step outside the manifold and move further away, then the movie would be that of images which initially look slightly like cats and gradually turn into random patterns and noise-like images. Note that parts of cats are themselves lower-dimensional manifolds, and the cat manifold has a compositional part-whole topology which should be exploited by AI in recognizing cats.

---

* Exercise 1   Image Manifold

Formally show that an image class is a manifold in raw pixel space. For an excellent book on manifolds, refer to [126].

Note on Exercises and Solutions For solutions to all the problems in the book, refer to Appendix A. Exercise 1 is typical of the exercises in the book. The underlying philosophy is to emphasize proactive learning, in which the student is encouraged to refer to the references. With the rapid proliferation of excellent online reference material, this book is intended to guide the learning of readers in directions, which build conceptual, geometrical, and intuitive understanding. Though some of the references have advanced material such as complex mathematical proofs, the exercise of looking them up enables the readers to build a holistic view of the field. The asterisks next to exercises indicate the level of difficulty.

1.2   Sculpting Process

For Rodin and Michelangelo, the sculpting process involves chiseling off the stone based on artistic insight. In AI, it is a mechanical process. We are given two sets of training points: (1) a positive set of points inside the final sculpture and (2) a negative set of points outside the sculpture. We are not told about the geometry of the actual sculpture; therefore, we can't really do a perfect job. The holy grail of AI is to create a sculpture which will contain all possible positive points with the negative set chopped off, irrespective of whether the points are training (seen) or test (unseen) ones. Then, the training error and the test error will both be zero, a highly unlikely situation for any practical problem.

The use of training data is the famed *training process* in machine learning. That's where the word *learning* comes in ML; the field is about making machines learn using data-driven methods. The way the learning works is as follows. We choose a machine learning model, which will determine up to what approximation we can create the sculpture. The AI will learn how to sculpt gradually by repeatedly making cuts, evaluating them, and improving them. Again and again, the AI will learn and improve till it gets it almost right.

Once the AI has created the final sculpture, it stops improving it. The learned model can then be put to practical use. It can then tell whether a point belongs to the sculpture or not. This manifold membership problem is called *inference*. At inference, the AI model will still compute the carving cuts in a process called the forward pass as they are needed to find out whether an unseen new point is contained within those cuts, but with no desire to improve the model. During training, the goal of the carving process is to discover the best cuts by computing how the carving can be improved further in a process called the backward pass. We evaluate the results using a loss function and strive to minimize it. The training process uses the forward pass as well as the backward pass.

For linear models, we are allowed exactly one big cut through the stone. On one side is the manifold and on the other side is its complement. Clearly that will be a poor approximation for complex manifolds. With a Support Vector Machine (SVM), we can smoothly carve out a nonlinear boundary, which is definitely an improvement, explaining why SVMs were popular in the days just before AI.

In AI, we choose a Deep Neural Network (DNN) with many layers of neurons. That's why the term *deep learning* is often used interchangeably with AI. AI is the umbrella term for many techniques in which deep learning plays a central role.

The number of layers is the depth of a neural network and the number of neurons in a layer is its width. It turns out that such an AI model can approximate the sculpture far more accurately than SVM and other classical methods, which will have to depend on sophisticated feature engineering in order to be competitive. So far classical methods have failed to outperform AI in competitions. The reason is that hand-engineered features have to be made more robust in many diverse situations, and it becomes harder to achieve this robustness in competition with AI. AI has stayed ahead by making use of ever-increasing computing power and data. Classical ML with its hand-engineered features hasn't been able to catch up with AI.

Is AI the same as deep learning? Deep learning used to be a small part of AI. With advances in the last decade, it has become a big part of AI. Of course, AI also employs other techniques such as reinforcement learning, statistical relational learning, and probabilistic graphical models. These techniques can be naturally integrated with deep learning to build complete systems, as we will see later in this book, and this will be one of the directions for further work in the future. In earlier days of AI, there used to be a lot of focus on various domain-specific search techniques, logical rules and inference, and statistical modeling. Rule-based expert systems are subsumed by deep learning as a special case. In a logical if-then-else decision rule, the cutting hyperplanes are orthogonal to some axis in the feature space and one works with the Boolean operations. AI models work with arbitrary

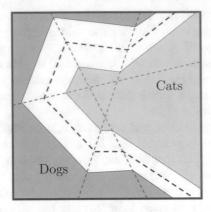

Fig. 1.3 AI chops off manifolds into convex polytopes. The dashed lines are the chiseling cuts. A cats versus dogs binary classification problem is illustrated here. In general, it could be cats versus not-cats. AI outputs a high probability of the input image being a cat (dog) in the green (red) region. It is not sure in the separating white margin. Every neuron contributes a chiseling cut. Neurons belonging to lower layers contribute simpler cuts. Higher-layer neurons contribute more complex cuts which bend more. An alternative and equivalent view is that the DNN is building a hierarchy of feature spaces by bending the input space in such a way that it becomes easier to separate out the classes in these feature spaces. Feature spaces built by the higher layers bend the input space more than those built by the lower layers

hyperplanes and can implement Boolean circuits, and most importantly, everything is being learned using data-driven methods.

The output of each neuron is followed by an activation function called a Rectified Linear Unit (ReLU). Conventionally, nonlinear activation functions, such as the hyperbolic tangent or the sigmoid functions, which are differentiable everywhere, were used. ReLU looks less smooth because it is not differentiable at zero, but it outperforms other activation functions. With the choice of ReLU, the sculpture gets approximated faster and in a piecewise manner. These pieces are convex polytopes akin to jigsaw puzzle pieces fitting together to solve the manifold puzzle. See Fig. 1.3.

AI does not directly chisel the surface of the sculpture and therefore it differs from human sculptors. It sculpts indirectly by cutting up the raw data space into numerous jigsaw pieces and then discarding those which are not part of the manifold, see Fig. 1.3. Every neuron contributes its cut. Neurons in lower layers make simpler cuts. Those in higher layers make more complex ones which bend around. With millions of neurons, it all adds up significantly. From the exterior of the manifold, it will look as if numerous flat faces have been carved.

The number of convex polytopes is the *expressive power* of AI. The higher the expressive power, the better the approximation. To calculate expressive power, consider a basic type of DNN, namely, a Feed-Forward Network (FFN); see Fig. 1.4.

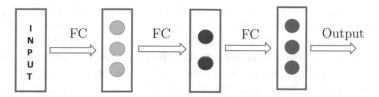

Fig. 1.4 Example of a feed-forward network (FFN). FC means *fully connected layer*, in which each neuron is connected to every neuron of the next layer. The input layer is a pass-through layer. In this example, there are a total of three processing layers, which include two hidden layers and one output layer. The width of the FFN is three and the depth is three

An FFN is often called a Multi-Layer Perceptron (MLP), which was the first type of neural networks studied, from the historical perspective. In a classical shallow neural network, there is only one hidden (intermediate) layer. We extend this to many hidden layers, which are fully connected to neighboring layers in a cascaded sequential manner. In a fully connected layer, a neuron is connected to every other neuron in the adjacent layer. The connections are weighted indicating how much neurons influence others to which they are connected. A neuron performs a weighted multiplication of its inputs, adds a bias term, and fires when the result is positive. The output is sent to neurons of the next layer. The weights and the bias term are the internal learnable parameters. Neurons fire in a cascaded manner and the computation flows from the input to the output.

In a DNN, the input layer is pass-through layer and does not perform any computation. Define a processing layer to be a hidden layer, including the output layer. Define the depth of an FFN to be the total number of processing layers. In general for any directed acyclic computation graph, the depth of a DNN is the length of the longest computation path from the input to the output. Define the width of an FFN to be the width of the widest processing layer. See Fig. 1.4.

*** Exercise 2   Expressive Power

Based on a literature survey, state how the number of convex polytopes increases as we make an FFN wider and deeper. In particular, refer to [145, 171, 253]. The primary goal of this exercise is to look up references and be able to state the main results of a research paper without going through the details of mathematical proofs, which can be fairly complex. For readers who wish to get intuition about the mathematics, we will count the number of convex polytopes later in Sect. 4.3 once we understand the geometrical carving process.

The above exercise shows that AI chisels the sculpture very finely, being given a much larger number of chisels compared with classical ML. This is truly amazing and explains partly why the AI revolution is happening. Of course, classical ML doesn't need such high expressive power if it is working in a carefully designed feature space, where the sculpture is easier to carve. The problem is in the design of such a feature space, which is a difficult undertaking. AI doesn't have to deal with this problem.

> *The AI model carves out a manifold very finely compared with classical ML.*
> *It is given a generous budget of convex polytopes to carve out the manifold*
> *which grows exponentially with its depth and polynomially with its width.*

1.3   Notational Convention

Before we continue further, a word on notational conventions. Some authors use notations, which help the readers distinguish between scalars (numbers) and vectors. For example, lower-case Greek symbols may be used for scalars and lower-case Roman letters for vectors. Other conventions for vectors include use of boldface font ($\mathbf{x}$) or of arrows over them ($\vec{x}$). In physics, overline notation ($\overline{x}$) is sometimes used. These notations are not standard and one should always use the context to figure out what things are despite the author's stated notations. In this book, we typically use lower-case letters for vectors unless specified otherwise in the context. The use of a subscript in x_i is to denote the i-th component of a vector x. At the same time, x may refer to a set of vectors at times and then x_i is the i-th vector in the set x. In the following paragraphs, where we write $F(x)$, it is clear from the context and the surrounding text that x is a vector. Some authors may have chosen $F(\mathbf{x})$ or $F(\vec{x})$. Later in this chapter, we write

$$\theta = (\theta_1, \theta_2, \ldots, \theta_k) \in \Theta,$$

where it is clear that θ is a vector, θ_i is a scalar component of θ, and Θ is a set of vectors.

1.4   Regression and Classification

In the preceding sculpting analogy, we are looking at classification. The output is a category, a class, or a label. In regression, the output is a value, typically a real number. Given an image of a cat, you need regression if you want to know its age or the distance between its eyes. You need classification if you want to know if it is a kitten or an adult. Another example of regression is prediction of bounding box coordinates enclosing the cat; this is called object detection in computer vision. Another example of classification is wanting to know whether an online product review by a customer is positive or negative; this is called sentiment classification. Regression and classification have a close relationship.

1. In regression, we are fitting a function. For example, linear regression is a widely studied classical ML example, where one is fitting a hyperplane. Given some predictor variables (features) x, the output is the predicted value of the target or response variable y, which depends linearly on the predictor variables,

$$\hat{y} = F(x),$$

   where x is the vector of predictor variables, $\hat{y}$ is the predicted response variable, and F is the fitted linear function.
   In classification we have manifolds and we can reduce the problem to that of regression by fitting to the characteristic function of a class, under a suitable loss function. For the cat manifold, its characteristic function will take value 1 for a point belonging to the manifold and 0 otherwise. In carving out the cat manifold, we assign high probability to points in the support of the characteristic function.
2. In classification, we have manifolds which we are carving out. The process of carving is a numerical process because the output is a probability value for every input sample,

$$p = F(x),$$

   where p is the predicted probability value of the point x belonging to the manifold. Our sculpting process allows us to shade the convex polytopes with values in the range [0, 1] for classification and in $\mathbb{R}$ for regression.
   If we want a scalar-valued output, it is like sculpting with shade in grayscale as in Fig. 1.5. If we want a 3-D vector as the output, this is analogous to painting convex polytopes with RGB colors. There will be three output neurons, each one corresponding to a color component.

Clearly, classification and regression are two sides of the same coin, where the coin is the carving process. They share the same carving process in AI.

1.4.1   Linear Regression and Logistic Regression

As a concrete example, consider the close relationship between linear regression and logistic regression. In linear regression, the fitted linear function itself is the final answer and the loss function is L^2 or L^1 error. The loss function is also known as the cost function or objective function.

Fig. 1.5 A linear function
fitted over a convex polytope
for regression. Squishing of
the fitted function to the [0, 1]
range solves classification
under the cross-entropy loss

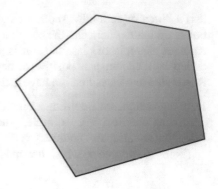

Linear models for classification do the same but squish the fitted function to the
range [0, 1], converting values into probabilities. This is achieved by the sigmoid
function for scalar-valued output and by the softmax function for vector-valued
output. The input space is split into two half-spaces by a cutting hyperplane. If
the fitted function is $f(x)$, then the hyperplane is given by the equation $f(x) = 0$.
One half-space corresponds to the positive class, where $f(x) > 0$ and the other
half-space corresponds to the negative class where $f(x) < 0$.

This is the method of logistic regression, which is *a classification technique
despite the misnomer*. The distance to the cutting hyperplane in the input space
translates into the probability of a sample x being in the positive class or the negative
class. The farther the sample is from the dividing hyperplane in the positive half-
space, the higher is the probability for the positive label. The same holds for the
negative label in the negative half-space. If x is right on the separating hyperplane,
then the probability is 0.5.

In logistic regression, the loss function is the cross-entropy loss. The cross-
entropy loss works nicely with probability distributions, and therefore it is routinely
used for classification.

In a nutshell, when going to classification from regression, the loss function
changes. And there is an extra squishing function at the end. This squishing is
performed by the sigmoid activation function of the output neuron. If there are
multiple output neurons for a multi-classification problem, then it is performed by
the softmax layer.

1.4.2   Regression Loss and Cross-Entropy Loss

In the world of loss functions, the L^2 loss for regression and the cross-entropy
loss for classification reign supreme. The L^1 loss also makes it to the top. It is
worth devoting a section to develop intuition about these loss functions. At the most
fundamental level, a loss function measures the discrepancy between the output of
an AI or ML model and the ground truth.

First, consider regression. Say, the training set has n input samples,

$$x_1, x_2, \ldots, x_n,$$

where each x_i is a vector of features, with the ground-truth values,

$$y_1, y_2, \ldots, y_n,$$

where each y_i is a scalar value, and the linear model predicts

$$\hat{y}_1, \hat{y}_2, \ldots, \hat{y}_n.$$

The error vector is

$$e = [y_1 - \hat{y}_1, y_2 - \hat{y}_2, \ldots, y_n - \hat{y}_n].$$

The regression loss is the length of the error vector. This measures the departure of the predicted values from the ground-truth values. We know the Euclidean length of a 2-D vector (v_1, v_2) is

$$\sqrt{v_1^2 + v_2^2}.$$

That's not the only way to measure the length. The Manhattan length is

$$|v_1| + |v_2|.$$

If you are asked to go 300 m west and 400 m north in Manhattan, New York, you will walk 700 m though the Euclidean distance is 500 m. The length of a vector is known as the norm of the vector. The same formulas work for n dimensions,

$$\sqrt{v_1^2 + v_2^2 + \ldots + v_n^2},$$

$$|v_1| + |v_2| + \ldots + |v_n|.$$

The L^2 loss is the Euclidean length (the L^2 norm) of the error vector and the L^1 loss is the Manhattan length (the L^1 norm). The former is more sensitive to outliers due to the square operation, which amplifies large values. In general, the L^p loss for regression, $p \geq 1$, is defined as

$$(|y_1 - \hat{y}_1|^p + \ldots + |y_n - \hat{y}_n|^p)^{\frac{1}{p}}.$$

For classification, suppose you are classifying cats versus dogs. The logistic regression outputs a probability value p indicating the likelihood of the input being a cat. If it is cat, ideally we want $p = 1$, and for a dog, $p = 0$. Suppose we are given five images,

$$x_1, x_2, x_3, x_4, x_5$$

where the respective ground-truth labels are

Cat, Cat, Dog, Cat, Dog.

Let the predicted probabilities of each image being a cat be

$$p_1, p_2, p_3, p_4, p_5.$$

We want p_1, p_2, p_4 to be high and p_3, p_5 to be low. In other words, we want the likelihood of observing the data to be high according to the model. For cat images, we maximize the likelihood

$$p_1 \times p_2 \times p_4$$

and for dog images, we minimize the likelihood

$$p_3 \times p_5.$$

In order to consolidate both goals into a single one, we maximize

$$p_1 \times p_2 \times (1 - p_3) \times p_4 \times (1 - p_5).$$

This is a product of small numbers. We modify it in two ways: (1) make it more amenable to computation, by converting it into a sum by taking the logarithm to the base 2, and (2) make it negative to turn it into a minimization problem

$$-\log p_1 - \log p_2 - \log(1 - p_3) - \log p_4 - \log(1 - p_5).$$

This is the negative log-likelihood of the observed training data, and it is known as the cross-entropy loss function. You may be wondering how cross-entropy relates to Shannon's information-theoretic entropy of an event with occurrence probability p,

$$-\log p$$

and of all the events on average,

$$-\sum p_i \log p_i,$$

where the logarithm is to the base 2. If an event probability p is high, it is a common event, with less information content and not much element of surprise. If p is low, it is a rare event, with more information content and a big element of surprise. Therefore, the cross-entropy loss function is saying that the probabilities should move in a direction where we want the likelihood of the positive samples to be high and that of the negative samples to be low.

The total regression or classification loss over a set of samples is the sum of losses over individual samples,

$$\mathcal{L}(\{x_1, x_2, \ldots, x_n\}) = \sum_{x_i} \mathcal{L}(x_i),$$

where $\mathcal{L}$ is the loss function.

** Exercise 3   Loss Functions

How does the mean squared error (MSE) relate to the L^2 error? What is the shortcoming of the L^2 loss? Write the cross-entropy loss in terms of the ground truth and the predicted probability values. Show its connection to the Kullback-Leibler divergence. It is easier to optimize a function if it is convex. Comment on the convexity of these loss functions for the linear models.

1.4.3 Sculpting with Shades

The idea behind linear regression and logistic regression holds for DNN. We use the same loss functions. The fitted function just happens to be far more complex. Each convex polytope carved by AI has a linear function fitted over it. Such a convex polytope is called a linear region; see Fig. 1.5.

The entire input space gets divided into convex polytopes and fitted with a piecewise linear function. We have K linear functions,

$$f_1(x), f_2(x), \ldots, f_K(x),$$

where K is the total number of convex polytopes and the final piecewise linear function is

$$f(x) = f_j(x),$$

if $x \in j$-th convex polytope. For regression, we are done at this point. The fitted function f is the answer.

For a classification problem, there is one additional step. Just like in the case of logistic regression for binary classification, we squish the fitted function to the range $[0, 1]$,

$$F(x) = \sigma(f(x)),$$

where σ is the sigmoid activation function. For the multi-class case with C categories, the output $f(x)$ is a C-dimensional vector, and we use the softmax function,

$$F(x) = \text{softmax}(f(x)).$$

The intuition behind the sigmoid function and its generalization to the softmax function is worth mentioning. Suppose you are given a vector,

$$f(x) = [-1, 0, +2],$$

which needs to be squished. Positive-valued weights are obtained by applying the exponential function to give

$$e^{-1}, e^0, e^{+2},$$

which are then normalized to a probability vector with values summing to 1, by dividing each term by the total sum,

$$p_i = (F(x))_i = \frac{\exp((f(x))_i)}{\sum_j \exp((f(x))_j)}.$$

The rationale behind using the normalized exponential is that it highlights the maximum value. Suppose we are classifying five types of animals and we are given an image of a cat. Suppose we get the following reading of the fitted piecewise linear function,

$$f(x) = [+10, -1, 0, +2, -3],$$

where the first component is for cats. The weight for the cat component will be e^{10}. Squishing the vector to the $[0, 1]$ range by the softmax function will imply that the probability of the input being a cat dominates the others,

$$P(\text{Cat}) \gg P(\text{other animal}).$$

That's why it is called softmax. To prevent overflow/underflow problems in numerical calculations of the sum of exponential terms, the LogSumExp function, which takes the logarithm of the sum, is often used.

> *An AI model assigns numerical values to each point in the raw data space. This fitted function is piecewise linear, where the pieces are convex polytopes. Interpretation of these as arbitrary values solves regression. Interpretation as probabilities after the additional squishing step solves classification.*

1.5   Discriminative and Generative AI

So far we have been discussing discriminative methods , which go from data to some useful attributes, such as class labels. Generative models work in the other direction: they go from attributes to data. They are appealing because they capture what an object truly is and what its most important characteristics are.

In discriminative methods, an approximation to the manifold is carved with our chiseling cuts. In generative models, our goal is to have a data-driven method, which will generate new points belonging to the manifold. Let us use basic mathematics to make the concepts clear. We use the usual notation,

$$f : A \to B$$

for a function f which maps the set A to the set B. The set A is the domain of f. The subset of B onto which the whole of A gets mapped is the range of f.

Let X be a set of objects of interest such as all 100×100 RGB 24-bit images of cats. A generative model for cats is a Turing computable function g such that its range is the set X and its domain is the set of all valid generative parameters Θ for cats. We may have intuition about some of these parameters, and often many will be unknown to us. In the latter case, they are known as latent variables. Both X and Θ are finite or countably infinite. If we endow Θ with a discrete probability distribution P, then we have a probability distribution over X:

$$g : \Theta \to X,$$

$$P(x) = \sum_{g(\theta)=x} P(\theta).$$

Often generative models are defined as those from which one can draw data samples x as per the data distribution $P(x)$. This is done by sampling $\theta \in \Theta$ and generating x. We defined the methods as Turing computable functions to avoid getting into uncountably infinite sets such as reals and continuous probability distributions over reals. And AI and ML are based on computation in the form of algorithms. Can one really assume that the set of cats in the universe is uncountably infinite? The Turing model can describe any method, including ad hoc customized, feature-based, or data-driven methods. And it reflects the working of the actual universe as per the Church-Turing thesis.

What is the definition of a generative parameter? Intuitively it is a characteristic of the objects of interest which is needed to describe them. We may know many of these based on our expertise and intuition. At the same time, many are not known to us, and therefore they are called latent parameters or latent variables. The above definition is about the existence of a Turing computable function, which is approximated by a data-driven method in AI and ML. It doesn't offer any clues about the construction of such a function on a latent parameter space. For generation of cats, we have two practical approaches.

Computer Graphical Approach   Here the Turing computable function is implemented by computer graphical algorithms. The generative parameter space is known to us, and each parameter has a semantic interpretation. We design the function based on our intuition and expertise.

Generative AI Approach   In AI, the Turing computable function is an AI model trained on a training dataset of images. The generative parameter space is unknown to us. The parameters are latent and may not have a direct semantic interpretation. We design the function based on large sets of images.

The above distinction is crucial in understanding generative AI. Interestingly some of the latent generative parameters, which are learned, have interpretable semantic meaning, such as how much a person is smiling or the orientation angle of the face. The same distinction applies between the following two approaches for discriminating between cats and dogs.

Hand-Engineered Approach   Here, the Turing computable function is implemented by hand-engineered features, which are input to a classical ML model or to a rule-based system. The system is partly data-driven and is largely based on human expertise and intuition.

Discriminative AI Approach   This is purely a data-driven method. The Turing computable function is an AI model trained on a training dataset of real-world examples. The internal features, which are computed by the AI model, are unknown to us. They are latent and may not have a direct semantic interpretation.

Again, interestingly it has been observed, for example, in the case of computer vision, that the first convolution layer corresponds to wavelet-like filters. For higher-layer features, there may be a semantic interpretation, such as the face of the cat. We can summarize the above as follows for computer vision.

	Hand-engineered	Data-driven
Prediction/classification	Features and rules	Discriminative AI
Generation/synthesis	Computer graphics	Generative AI

We contrast the generative model with the discriminative cats versus dogs problem. In a generalized image understanding problem, we seek to build a function, which maps a cat or a dog image to its class label and to its generative parameters (also known as attributes, features, or properties),

$$f : C \cup D \to G \times \Theta,$$

where C is the cat manifold and D is the dog manifold. G is the set of labels,

$$G = \{\text{cat}, \text{dog}\},$$

and Θ is the set of vectors of valid k generative parameters describing age, size, texture descriptors, distance between eyes, location, orientation, length of tail, etc.,

$$\theta = (\theta_1, \theta_2, \ldots, \theta_k) \in \Theta.$$

The function f is performing regression as well as classification. When we care just about the class label, which is a categorical generative parameter, then

$$f : C \cup D \to G,$$

and the problem gets reduced to the standard classification problem. We can modify the categorical labels of interest; for example, if we include a third class, which is the complement of $C \cup D$,

$$f : \mathbb{R}^n \to G \cup \{\text{neither}\},$$

then the input can be any image and AI will carve out three manifolds. For the cats versus not-cats problem, we generalize the dog class to the set of all negative examples,

$$G = \{\text{cat, notcat}\}.$$

Both generative models and discriminative models are Turing computable functions in their most general definition. The difference between the two is based on our semantic interpretation of the input and the output. It is the difference between computer graphics and computer vision. The former generates an image based on the given model parameters, and the latter infers model parameters when given an image. A generative model will write a story, paint a painting, or compose a song. A writer or an artist is a generative model and produces content. Starting with the desired attributes of a cat, an artist paints a cat. The input is a model and the output is an image. A discriminative model will take a story as the input and determine if it is sad, happy, or neutral. It will take a painting and decide whether it is by Monet or Vincent van Gogh. A censor officer is a discriminative model and judges whether to censor the content or not. A judge is a discriminative model who categorizes, evaluates, predicts, and decides, when presented with details (data) of a case. The input is a data sample and the output is a decision. See Fig. 1.6.

In AI, we consider a very specific class of Turing computable functions, which are discovered using data-driven methods trained on the raw data space. Both discriminative AI and generative AI will learn to perform their respective tasks based on the given training set. For the former, we need both the positive set and the negative set. For the latter, we just need the positive set. The function f to be learned is parameterized by the learnable parameters internal to the AI model. The architecture of a DNN implementing f is fixed in advance, and the values of the trainable weights and bias terms are learned.

Fig. 1.6 Judge versus artist. Discriminative AI versus generative AI. The former evaluates, decides, censors, and categorizes. The latter generates content. For input-attribute pairs (x, θ), discriminative AI models the conditional probability distribution $P(\theta|x)$ and generative AI models the joint distribution $P(x, \theta)$

> *In discriminative methods, we are testing membership of an unseen point in a manifold. In generative methods, we are generating unseen points belonging to a manifold.*

1.6   Success of Discriminative Methods

For discriminative models, we have recently found tricks, which solve the problem for practical applications. With this paradigm, training is slow and inference fast. Training takes time, whereas the fitted function is a quick shortcut to get an approximate answer. We don't care to understand how the object was generated, and we are pragmatically keen to build products. We want speedy inference and we are willing to invest resources in data collection and in training.

For generative models, research is under progress. Interestingly, just like state-of-the-art discriminative models, generative models are based on large amounts of data. In discriminative models, it is easy to define the input and the output and then fit a function to the given input-output pairs. In generative models, it is not even clear what the input should be. What generates a cat? It is a difficult question.

Therefore, we have had greater success on the discriminative side. At the same time, new innovative research is making significant inroads into the challenges of generative AI. We will see that the line between generative AI and discriminative AI is not clear. One can develop approaches for discriminative tasks based on generative AI. And both use the same carving process because both are implemented by deep neural networks. It is our semantic interpretation of the input and output spaces, which makes them look different.

1.7 Feature Engineering in Classical ML

Discriminative ML models have drawn inspiration from generative models. Though we don't know what all generative attributes of a cat are, we can apply our insight and come up with a candidate list and implement those. And we can strive hard to discover those attributes, which distinguish cats from dogs and from other negative examples. These attributes are called features when applied to the task of classification or regression, and the process is called feature engineering; see Fig. 1.7.

Features are painstakingly computed based on human expertise, experience, intuition, wisdom, and labor. These are salient characteristics of cats such as the presence of whiskers, eyes, ears, nose, fur, paws, tail, and others, which the machine learning engineer considers to be important for the task at hand. Advanced computer vision and mathematical algorithms are used, and great care is taken to make them invariant to a cat's size, pose, and orientation. Examples of human-engineered features in computer vision are Histogram of Oriented Gradients (HOG) and Scale-Invariant Feature Transform (SIFT). Once D many features have been extracted, cats become points in the D-dimensional feature space.

A feature space will typically have dimensionality less than that of the raw pixel space. Classes in a feature space are considered easier to carve out than the raw data space. The primary goal of feature engineering is to make classes easily separable in the feature space. AI showed us in 2012 that we can do the same job in the raw data space. This was the famed ImageNet competition in which a DNN working in the raw pixel space was able to classify one thousand different types of animals, birds, things, and objects, outperforming the classical ML competitors.

Once we have features, we move from the raw data space to the feature space. The cat manifold becomes a region in this feature space, so does the dog manifold. Will these regions be manifolds? If the feature extraction constitutes a homeomorphism, then the manifolds in the original pixel space will be mapped to manifolds in the feature space. However, there is no guarantee that feature extraction preserves the manifold property in general. We will say that the manifolds in the input space have been mapped to sets in the feature space.

Fig. 1.7 Manual feature engineering in classical ML. AI bypasses it as shown by the red arrow. AI performs automated feature engineering

> ### * Exercise 4   Feature Space
>
> What property will feature need to satisfy so that an image class is a manifold in the feature space?

The set of features is often pruned by selecting a subset of useful features using statistical techniques and discarding those which are noisy. The dimensionality of a feature space is further reduced by using a dimensionality reduction technique to make it more tractable. Therefore, the classical discriminative ML pipeline is:

1. Collect training data.
2. Extract useful features. This gives you a feature space in which it is easier to separate out the classes.
3. Optionally, reduce the dimensionality of the feature space.
4. Carve out the classes in the feature space using ML.

The breakthrough which deep learning has provided is that we can do away with the middle two steps. These two omitted steps get subsumed implicitly by AI, and we work directly with manifolds rather than sets or regions in the feature space:

1. Collect lots of training data.
2. Carve out the manifolds using DNN.

Internally, AI is performing feature extraction and dimensionality reduction steps in its various layers. They are just opaque to us, hidden inside the black box of AI models. AI has automated these steps for us, and therefore AI is often viewed as an automated feature engineering technique, reducing the burden on ML scientists. In fact, AI goes one step further. We will see that it builds a hierarchical representation of the input in terms of these automatically computed features and that it is a powerful representation learning technique. An input gets represented by the activation pattern of neurons. This is called a *distributed representation* of the input. An input gets mapped to a point in a latent semantic space. Inputs, which belong to the same class, have similar representations, and therefore they can get separated out from other classes.

Can one perform deep carving of a feature space instead of a raw data space? Yes, if the feature space is very high-dimensional, then AI computes "features of features" and the same expressive power can be employed using a multilayer network. After all, the raw data space can be viewed as the most fundamental feature space.

AI carves out manifolds in the raw data space in contrast to classical ML, which carves out regions in a feature space. Somewhere in the intermediate stages of the carving process, AI is automating the feature engineering process in a hierarchical manner. It builds a distributed representation of the input in a latent feature space.

1.8 Supervised and Unsupervised AI

It is worth noting that pretty much all modern AI and ML are driven by data, especially annotated, labeled data. There is an exception. A distinction is often made between supervised ML and unsupervised ML, because labeled data is not needed in the latter.

In the supervised case, we are given the desired output, and there is a loss function to evaluate how well we are producing the right output. The target output is the result of the data annotation or labeling effort. Sometimes, labels can be implicit and can be extracted from the data, as in *self-supervised ML*. At other times, we can algorithmically extend labels from the annotated data to the unlabeled data, resulting in *semi-supervised ML*.

In the unsupervised case, we are interested in learning the geometry of data without being told anything else. It includes methods to find naturally occurring clusters, to reduce dimensionality, to explore and visualize data, and to detect anomalies and outliers. See Fig. 1.8 for an illustrative example. A fundamental question in unsupervised learning is as follows: What is the geometry of the cloud of points we are looking at? Is there a pattern there?

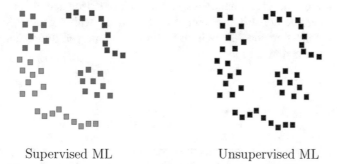

Supervised ML   Unsupervised ML

Fig. 1.8 Supervised versus unsupervised methods. We are given three classes (red, green, blue) for the supervised case, and the goal is to build a classifier. In the unsupervised case, we don't know the class labels, and the goal is to explore the geometry of the unlabeled dataset. We can visually infer that there are four clusters, two of which have one-dimensional geometry

> *Unsupervised AI and ML seek to understand the geometry of and to discover*
> *patterns in the given unlabeled data.*

1.9   Beyond Manifolds

This chapter was written to give readers geometrical intuition into AI and ML
from a bird's-eye viewpoint. We extensively used the concept of manifolds, which
works beautifully well for computer vision. AI has found many applications besides
computer vision [121], and there will be domains where you may not be convinced
of the manifold hypothesis. Perhaps there are temporal dynamics in the problem, and
we don't have a fixed manifold but a variable one which changes over time. Perhaps
the underlying geometry is a collection of arbitrary sets. At the extreme, it could
be a set of disconnected points. Looking at the complete picture, the entire system
will likely have many components, each with a different geometry, and therefore
the overall geometry will become less tractable as you combine the components
together. Extend Exercise 5 to explore the underlying geometry of the specific
problems of interest to you.

> ### ** Exercise 5   Data Manifolds
>
> Discuss whether typical classification problems in speech and audio, natural
> language understanding, and malware detection lead to manifolds in their
> respective raw data spaces.

Despite these conceptual challenges, if you are using deep neural networks in the
raw data space, you are still approximating a solution with a geometrical carving
process. The following chapters will take you on a journey to understand the inner
dynamics of the sculpting process.

1.10   Chapter Summary

A strong geometrical intuition into the working of AI can be developed with an
analogy with the sculpting process. For classification, an AI model sculpts a high-
dimensional manifold with exponentially many convex polytopes. For regression, it
fits a piecewise linear function with exponentially many pieces (linear regions). This
exponential expressive power enables AI models to solve challenging problems. The
training process teaches an AI model how to sculpt based on a training dataset and

a loss function. For regression, an example of a loss function is the L^2 loss, and for classification, the cross-entropy loss is widely used. The L^2 loss measures the length of the error vector, and the cross-entropy loss is based on the likelihood of data. Linear regression and logistic regression are two linear models in classical ML, and many important concepts in ML can be studied in the context of linear models. AI models can be divided into two major kinds: (1) generative and (2) discriminative. Generative AI is able to create new unseen points in a manifold. In their full generality, both types of AI are Turing computable functions, which are approximated by data-driven methods. A connection to the theory of computation is needed to emphasize that AI works with countably infinite sets, rather than uncountably infinite sets such as reals. Another distinction is between supervised methods and unsupervised methods. The latter seek to learn the geometry of the given data and to find natural patterns in it. The key difference between classical ML and AI is that the former works in carefully hand-designed feature spaces whereas the latter works in the raw data space. AI performs automated feature engineering and builds a distributed representation of the input in terms of hierarchical features.

Problem Set 1

1. Give an example of a 1-D set which is a manifold. Given an example of a 1-D set which is not a manifold. Is the earth's surface a manifold? Is the whole earth a manifold?
2. Consider a point in the cat manifold which corresponds to a cat. Suppose the cat is sitting still against an unchanging background. A person enters the background and leaves. What would this time sequence correspond to in the manifold? Now consider that the background remains static and the cat slowly gets up and leaves the scene. What would this time sequence correspond to in the manifold?
3. In a very high-dimensional space, where is most of the volume of a sphere located? What is your intuition behind the geometry of manifolds of common image classes in such spaces?
4. Define the unit circle as the set of points at a unit distance from the origin. In the L^2 norm case, the unit circle is the standard circle of diameter of two units. What is the unit circle in the L^1 norm case? For L^∞?
5. Consider a probability space of two outcomes with the uniform probability distribution of [0.5,0.5]. Compute the average entropy. Then, recompute it for the distribution [0.1,0.9]. Derive the probability distribution which has the maximum entropy.
6. What is the softmax of $[x, x]$ for any $x \in \mathbb{R}$. What is the difference in the softmax of $[x, y]$ and of $[x + z, y + z]$ for any $z \in \mathbb{R}$?
7. Suppose you are designing a classical computer vision algorithm to distinguish images of natural landscapes from those of city street scenes. Which of the following will be good features: (1) straight lines, (2) color histograms, (3) beauty of scenery, (4) serenity of scenery, (5) text regions, (6) people, (7) vehicles, (8) expressions on people's faces, or (9) time at which the photo was

taken? Suppose you come up with a tentative list of 200 features for this task. Is it easy to select a subset of these features which will lead to the best result?

8. According to the theory of computation, can an AL model ever work with real numbers? What is the difference between an AI model and a Turing machine?

9. On his/her own, a child notices that his/her dog has four different types of facial expressions. The child then practices to sketch the dog. Later the child goes to his/her school, and the teacher at the school teaches different letters of the alphabet. What type of learning is the child performing?

10. How will you make image features invariant to scaling and rotation in classical machine learning?

Chapter 2
Make Me Learn

> You don't have to see the whole staircase, just take the first step.
>
> Martin Luther King, Jr.

> It is a capital mistake to theorize before one has data.
>
> Sherlock Holmes

When you visit the Grand Canyon National Park, you are filled with wonder the moment you view it from the rim. It is truly a grand spectacle created by nature. If you are an adventurous, experienced hiker, you can embark on a downward trip toward the rapidly flowing Colorado River by following one of the well laid-out paths. You take the first step, then the second one, and you are on your way.

AI learns in a similar way. It faces a challenging and unknown landscape, and the goal is to reach spots where the loss function is very low, just like the Colorado River Valley. Unlike human hikers, it doesn't have a path laid out and it has to discover it. This is exactly what *learning* is in ML. In this chapter, we get a glimpse of how it happens. It occurs because starting from an initial spot, an AI learning process takes baby steps toward its goal all guided by the data. These steps are known as gradient descent steps. This journey of billions of steps takes place in a surreal high-dimensional world called an optimization landscape, also known as a loss landscape or loss surface. We will build intuition into how these steps are found and how this journey takes place.

S. Dube, *An Intuitive Exploration of Artificial Intelligence*,
https://doi.org/10.1007/978-3-030-68624-6_2

We make a crucial observation at this point. Algorithms play a central role in computer science, and they have been catapulted to a prominent position in the world with the rise of technology based on algorithms. Study of algorithms has been an intellectually stimulating field, and students of computer science learn famous algorithms such as quicksort and heapsort for sorting, Kruskal's and Prim's algorithms for minimum spanning trees, and Dijkstra's and Floyd-Warshall's algorithms for shortest paths in graphs. Software is written to implement specialized algorithms in order to solve practical problems, and AI now shows a new way of designing algorithms and of programming. In this new style, we specify an AI model and make it learn using data. There is no need to have hand-engineered techniques and highly customized solutions because the same general principles can be applied to several problems of practical significance across different domains.

It is like having general hiking techniques which allow humans to not only hike down the Grand Canyon in Arizona but will enable them to explore the Valles Marineris canyon on Mars in the future.

2.1 Learnable Parameters

2.1.1 The Power of a Single Neuron

To understand the training process, we start with a single neuron. The most fundamental unit of human brains as well as of artificial neural networks is a neuron. A biological neuron consists of dendrites and an axon. Dendrites receive impulses from other neurons, and the axon transmits the impulse to other neurons. If the net strength of incoming excitatory signals is over that of incoming inhibitory signals by a threshold, the biological neuron fires and transmits the impulse.

The artificial neuron is inspired by the biological neuron; see Fig. 2.1. For an input vector $x = (x_1, x_2, \ldots, x_n)$, it implements a mathematical function:

$$y = f(x) = \sigma \left(b + \sum_{i=1}^{n} w_i x_i \right),$$

where w and b, the weights and the bias term, respectively, are $(n + 1)$ learnable, trainable, or adjustable parameters. The function σ is called the activation function of the neuron because it determines how the neuron will get activated. The computed value y is called the activation or the output value of the neuron. There are three common choices for the activation function:

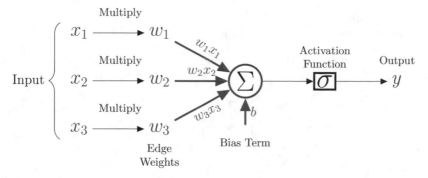

Fig. 2.1 Artificial neuron. With proper choices of the activation function, it can implement the linear models of classical ML

- The hyperbolic tangent function $f(z) = (e^z - e^{-z})/(e^z + e^{-z})$.
- The sigmoid function $f(z) = 1/(1 + e^{-z})$.
- The ReLU function $f(z) = \max\{0, z\}$.

A single neuron is capable of solving ML problems. Here are some different scenarios:

- If σ is the identity function and the parameters are trained under a regression loss function such as the L^2 loss, then a single neuron implements linear regression.
- If σ is the sigmoid function and the parameters are trained under a classification loss function such as the cross-entropy loss, then a single neuron implements logistic regression. The sigmoid function squishes the output to the range [0, 1].
- If σ is the ReLU function, then it implements a ReLU neuron, which allows the positive output to pass through while a negative output is squashed to zero. The neuron is called a ReLU neuron.

ReLU neurons are widely used for hidden layers in modern AI models because they lead to faster convergence of the training process. For classification, the sigmoid activation function or the softmax function has to be employed at the output layer.

2.1.2 Neurons Working Together

Consider the simplest scheme of cascading several fully connected hidden layers in an FFN; see Fig. 1.4. We first open the black box to peer inside an FFN with one hidden layer in Fig. 2.2.

In an AI model, there are several hidden layers with a large number of neurons. For Fig. 2.2, the set of learnable parameters consists of the parameters of the neurons

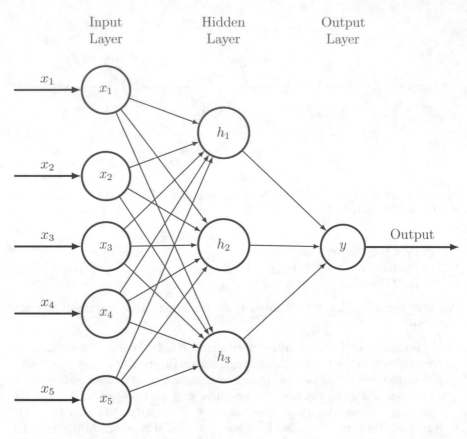

Fig. 2.2 Detailed view of an FFN with one hidden layer. Neurons are depicted as blue circles. The output of a neuron is shown inside its circle. See Fig. 2.1 to zoom in on a neuron. The hidden layer is fully connected to the input pass-through layer. In total there are 22 learnable parameters. Each hidden neuron has five incoming edges; therefore, it has five weights and one bias term. The output neuron has three incoming edges and one bias term. In practice, a DNN will have millions of parameters to be learned, and the training is done by taking many gradient descent steps. Each step uses the backpropagation algorithm, which involves a forward pass and a backward pass, to compute gradients of the loss function for a minibatch of training examples with respect to the learnable parameters. The learnable parameters are then tweaked after computing the gradients. The process repeats by taking the next step

in its hidden layer and the output layer. Let the number of neurons in an FFN excluding the input layer be N, and let the number of edges in the computation graph of the FFN be E. Then, the total number of learnable (adjustable) parameters is

$$E + N,$$

which includes the weights and the bias terms, respectively. There is a natural, intuitive mechanism of how the ReLU-based FFN works. The weights on the edges quantify the strength of the connections between neurons. The higher the magnitude of the weight, the greater the strength. A neuron takes the weighted combination of its inputs and "fires" if the net input signal is over the threshold $-b$, where b is its bias term, and transmits its output forward. Starting from the first hidden layer, the neurons fire in a cascaded fashion, transmitting their output forward, all the way to the output layer.

The total number of learnable parameters can be in the tens or hundreds of millions. The goal of training is to find those values of the parameters for which the loss function is minimized over the training set.

> *In a ReLU AI model, neurons fire in a cascaded manner as the computation flows from the input toward the output. Weights on the edges indicate the strength of connections between neurons. Either a ReLU neuron fires and transmits its output forward or it doesn't and blocks the transmission.*

2.2 Backpropagation of Gradients

One algorithm which revolutionized research on neural networks in the twentieth century was the backpropagation algorithm. In its essence, it is a smart way to implement the chain rule of taking derivatives in calculus for nested compositions of functions implemented by neurons. For seminal work on backpropagation, see [83, 182]. In Figs. 1.4 and 2.2, an FFN computes a function $f(x)$ which is a composition of functions computed by its neurons. The final output of the neural network is input to the loss function, and its result needs to be differentiated with respect to the learnable parameters:

$$\mathcal{L}(\theta, x) = \sum_{i=1}^{n} \mathcal{L}(\theta, x_i),$$

where x is a minibatch of n training samples and θ consists of the learnable parameters. In Fig. 2.2, θ has 22 parameters. For the (squared) L^2 loss, the expression is

$$\mathcal{L}(\theta, x) = \sum_{i=1}^{n} (y_i - f(x_i))^2,$$

and for the cross-entropy loss,

$$\mathcal{L}(\theta, x) = \sum_{i=1}^{n} -y_i \log f(x_i) - (1 - y_i) \log(1 - f(x_i)),$$

where for the input x_i, y_i is the ground-truth label and $f(x_i)$ is the output of the network. Let θ have k learnable parameters:

$$\theta = (\theta_1, \theta_2, \ldots, \theta_k).$$

Then the derivative of the loss function with respect to θ is a k-dimensional vector.

2.2.1 Partial Derivatives

In calculus a function is written as

$$f(x_1, x_2, \ldots, x_n),$$

where the x_i's are called variables. The function depends on the variables, and one can differentiate the function with respect to them. In the case of an AI model, the loss function depends on the following:

Parameter terms   These are the weights on the edges and the bias terms of neurons, and these are the variables which need to be tweaked. They have values assigned to them which have to be updated to newly computed values. Therefore the primary goal is to differentiate the loss function with respect to these terms.

Input terms   These are fixed and serve as constants. At the same time, there is no reason why we can't make them learnable in some applications. We will see later that in visualization of neurons and in DeepDream, the input terms are made variable and a suitably defined function is differentiated with respect to them.

Intermediate terms   These are the output values of all intermediate operations and of neurons, and they depend on the input terms, the parameter terms, and the preceding intermediate terms. These are intermediate variables in the underlying computation graph, and the loss function is differentiated with respect to these terms in order to implement the chain rule.

The partial derivative notation ∂ is used in multivariate calculus. Given a multivariate function such as $f(x_1, x_2)$, its partial derivatives with respect to its input variables are given by

$$\left(\frac{\partial f}{\partial x_1}, \frac{\partial f}{\partial x_2}\right).$$

The backpropagation algorithm computes efficiently the derivative of the loss function with respect to k learnable parameters:

$$\frac{\partial \mathcal{L}}{\partial \theta} = \left(\frac{\partial \mathcal{L}}{\partial \theta_1}, \ldots, \frac{\partial \mathcal{L}}{\partial \theta_k}\right).$$

2.2.2 Forward and Backward Passes

The backpropagation algorithm is divided into two passes.

Forward Pass   Assign values to all the input terms. Run the computation from the input all the way to the output, firing the neurons in a cascaded fashion and computing values of the intermediate terms. Extend the computation by one step further to the loss function, which is based on the output value and the ground-truth value. Note that as a result of the forward pass, every term gets a numerical value assigned to it.

Backward Pass   Start computing the derivative of the loss function with respect to the output variable. Run backward through the computation graph from the output toward the input computing partial derivatives along the way with respect to the parameter terms and the intermediate terms as you encounter them. Use the chain rule of differentiation from high-school calculus. Compute numerical values of the partial derivative expression terms. Note that as a result of the backward pass, each partial derivative term gets a numerical value.

In high school, students go from expression to expression in calculus using symbolic differentiation. Thus the derivative of x^2 is $2x$ and of $\sin 2x$ is $2\cos 2x$. In the backpropagation algorithm, we go from values to values because we are evaluating these expressions. If $x = 3$ in the high-school example of x^2, we will get the forward pass value of 9 and the derivative value of 6. If $x = \pi/6$ in the $\sin 2x$ example, then the forward pass gives $\sqrt{3}/2$, and the backward pass gives derivative 1.

Let's take another example. Let $x = 2$ be the input, and suppose there are four learnable parameters $w_1 = 3$, $w_2 = 1$, $w_3 = -5$, and $w_4 = 4$. Consider the following steps:

$$r = w_1 x,$$

$$s = w_2 r + w_3,$$

$$t = \text{ReLU}(s),$$

$$y = w_4 t,$$

$$L = (y - g)^2,$$

where r, s, and t are the intermediate terms, y is the output, and L is the L^2 loss with the ground-truth $g = 5$. The forward pass is

$$r = 6,$$
$$s = 1,$$
$$t = 1,$$
$$y = 4,$$
$$L = 1,$$

and the backward pass is

$$\frac{\partial L}{\partial y} = 2(y - 5) = -2,$$

$$\frac{\partial L}{\partial w_4} = \frac{\partial L}{\partial y}\frac{\partial y}{\partial w_4} = -2,$$

$$\frac{\partial L}{\partial w_3} = \frac{\partial L}{\partial y}\frac{\partial y}{\partial t}\frac{\partial t}{\partial s}\frac{\partial s}{\partial w_3} = -8,$$

$$\frac{\partial L}{\partial w_2} = \frac{\partial L}{\partial y}\frac{\partial y}{\partial t}\frac{\partial t}{\partial s}\frac{\partial s}{\partial w_2} = -48,$$

$$\frac{\partial L}{\partial w_1} = \frac{\partial L}{\partial y}\frac{\partial y}{\partial t}\frac{\partial t}{\partial s}\frac{\partial s}{\partial r}\frac{\partial r}{\partial w_1} = -16,$$

where we can save the partial common products in memory as soon as we compute them in order to reuse them in subsequent steps. Therefore, backpropagation is a dynamic programming algorithm for implementing the chain rule for graph-based computation. Note that if $x = 1$, then the ReLU will be inactive, disconnecting the graph, and making t and y zero and independent of the learnable parameters. The derivatives of the loss function with respect to the parameters will become zero. Only when the ReLU neuron wakes up with s turning positive will the dependence get re-established. One will have to wait for an appropriate input value of x for that to happen. What if such an input never arrives? This is the problem of forever asleep neurons.

*** **Exercise 6**   **Training**

1. Consider the following operations in logistic regression:

$$y = w_1 x_1 + w_2 x_2 + w_3$$

$$p = \text{sigmoid}(y) = 1/(1 + \exp(-y))$$

$$L = -G \log(p) - (1 - G) \log(1 - p)$$

Compute the partial derivative of L with respect to y. Assume the natural logarithm.

2. Can a neuron go to sleep and never wake up?

3. Consider the classification loss for K multiple classes. The ground-truth vector G is a one-hot vector. The loss for a training sample is

$$-\sum_{i=1}^{K} G_i \log p_i,$$

where $G_j = 1$ for the correct (positive class) j of the given training sample and $G_i = 0$, for $i \neq j$, elsewhere for the $K - 1$ negative classes, and p_i is the predicted class probability for class i. Therefore, all terms when $i \neq j$ become 0 and the total loss becomes

$$-\log p_j.$$

Does this mean that the loss is independent of what the AI model predicts for the negative classes?

Forward pass computes the composition of the functions:

$$y = f_N(\ldots f_1(x) \ldots),$$

by going in the forward direction from f_1 to f_N and by saving the intermediate terms,

$$z_k = f_k(\ldots f_1(x) \ldots).$$

The backward pass goes in the backward direction from f_N to f_1, differentiating according to the chain rule and making use of the intermediate terms:

$$\frac{\partial y}{\partial x} = \frac{\partial y}{\partial z_{N-1}} \frac{\partial z_{N-1}}{\partial x},$$

$$= \frac{\partial y}{\partial z_{N-1}} \frac{\partial z_{N-1}}{\partial z_{N-2}} \frac{\partial z_{N-2}}{\partial x},$$

$$= \frac{\partial y}{\partial z_{N-1}} \frac{\partial z_{N-1}}{\partial z_{N-2}} \cdots \frac{\partial z_1}{\partial x},$$

in which the products of common partial derivatives along the backward paths in the computation graph are saved for subsequent steps for efficient computation. Local partial derivatives are with respect to the local learnable parameters (variables) at each graph node, and they get multiplied by the saved products of the previously computed partial derivatives. This is like breadth-first graph traversal in the backward direction, computing local derivatives and making the products of the derivatives flow to the preceding nodes.

The reader must be asking why we can't work with expressions like in high-school calculus exercises, work out derivative expressions, and just plug in the values. The answer is that it is indeed possible but computationally intractable to derive an explicit expression for $\mathcal{L}(x, \theta)$ where x and θ are very large vectors. For AI models, it will be a complicated expression. The good news is that the expression is induced by a graph and therefore has a special structure amenable to computation. In backpropagation, one works with a graph-induced expression in an implicit manner by following the graph structure and reusing the intermediate computations. Still the question of directly working with the underlying expression as in high-school calculus is a deep one, and we will examine it further in Chap. 4 because it throws light on how AI works.

An intuition about the partial derivatives is as follows. The gradient g_w of the loss function with respect to a parameter w tells us that if we were to increase the parameter by an infinitesimal amount dw, then there will be a ripple effect down the computation graph, and the loss function will change by $g_w dw$. Therefore, an easy way to check if backpropagation is working fine is by perturbing w by a tiny amount and checking whether the loss function changes by approximately the expected amount. This is called gradient checking.

> *The backpropagation algorithm exploits graph-induced dependence of the loss function on learnable parameters to compute the gradients efficiently.*

2.3 Stochastic Gradient Descent

There is a price we have to pay for the tremendous expressive power of AI. Loss functions such as the L^2 loss or the cross-entropy loss which are convex for linear models become non-convex for AI models.

Let's first define a convex set. A convex set is one in which if one were to draw a line between any two points, then the line segment will be inside the set. In the US flag, a stripe is convex and a star is non-convex. There are no depressions in the perimeter of a convex set. Now we can define a convex function. If you look at the graph of a convex function, the region above the graph is a convex set. The presence of disconnected local minima means the graph has depressions and it is non-convex.

A convex function is easy to optimize as there are no local minima worse than the global minimum. The function is bowl shaped, and you can slide down toward any local minimum. It can have multiple minima, but they can't form a disconnected set. In fact, they all together form a convex set. A convex function is bowl shaped with either a flat convex bottom or one unique minimum. Convex loss functions for linear models and SVMs in the classical world can be optimized easily. We have to leave the world of convex optimization if we seek AI. The good news is that there is a training algorithm which is able to deal with the challenges of non-convex optimization.

Suppose we have a minibatch of n training samples and there are two learnable parameters w_1 and w_2. Compute the loss function for the minibatch over the 2-D parameter space. See Fig. 2.3 for an optimization landscape over such a parameter space. The vertical axis is the loss function. The horizontal 2-D plane is the learnable parameter space. We can see hills, valleys, and relatively flat areas. The goal is to reach a global minimum or a near-optimal local minimum.

Fig. 2.3 The optimization landscape (loss surface) of a loss function where there are two learnable parameters. SGD computes gradients of the loss function with respect to the two parameters. The gradient is a 2-D vector on the contour plot, and it will be perpendicular to the contour at that point

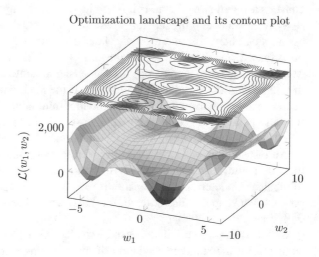

Optimization landscape and its contour plot

Imagine that we are standing somewhere in this landscape. We compute gradients using backpropagation for each sample in the minibatch. The gradients for the n samples in the minibatch are averaged. The result is an averaged 2-D gradient vector on the horizontal plane:

$$\frac{\partial \mathcal{L}}{\partial \theta} = \left(\frac{\partial \mathcal{L}}{\partial w_1}, \frac{\partial \mathcal{L}}{\partial w_2} \right).$$

It will be perpendicular to the contour line where we are standing.

A partial derivative is the ratio of infinitesimal change in the loss function when we turn the knob of an adjustable parameter by an infinitesimal amount. It doesn't give the size of the step we need to take. Therefore, we multiply it by a hyperparameter known as the learning rate $\alpha > 0$, typically chosen to be a small number with trial and error. We walk in the direction which lowers the loss function. The parameter values get updated:

$$w_1 = w_1 - \alpha \frac{\partial \mathcal{L}}{\partial w_1},$$

$$w_2 = w_2 - \alpha \frac{\partial \mathcal{L}}{\partial w_2}.$$

Note that we walk toward the valleys rather than the summits of hills. That's why there is a minus sign. We go in the opposite direction of the gradient because we want to minimize the loss function. If we turn the knob of a weight upward and the loss function increases, making the gradient positive, then we need to turn the knob downward. And if the loss decreases, making the gradient negative, that's the right adjustment direction and we turn it upward. If the learning rate is too small, convergence will slow down. If it is too high, then it may not converge at all.

Why do we take a bigger step when the gradient is bigger? There is an implicit assumption about the nature of the landscape around a local minimum that it is bowl shaped, and therefore when one gets closer to the minimum, then the gradient decreases. One takes larger steps for a steep descent into the valley and then reduces the step size to converge to the minimum. If the landscape was pathologically very flat, with a deep narrow depression around a minimum, then this strategy will fail. It is also possible to create SGD variants which normalize the gradient to the unit length so that they depend only on its direction.

And as soon as we have taken the step, a new minibatch arrives, and we find that the optimization landscape has slightly changed because it depends on the minibatch. Undeterred by this fluidity of the landscape, we continue the same process. We keep on sliding down the hills, which keep on shifting and morphing with time.

Should we choose minibatches to be small or big? The statistical sampling theory suggests that the standard deviation of the sampling error in the estimation of the true gradient will go down with a $\sqrt{n}$ factor, where n is the minibatch size. We should then feel more confident in taking larger gradient steps. The learning rate can be scaled up by a $\sqrt{n}$ factor; however, experiments suggest that it should instead be scaled linearly for good results; see [63].

There are finitely many minibatches and AI will eventually run out of them. Training data with N samples gives $\lceil N/n \rceil$ minibatches, where n is the minibatch size. When a pass over the entire dataset is over, we say that an epoch is over. AI is persistent and it wants to learn more. We shuffle the data and recreate new minibatches. The whole process repeats.

Epoch after epoch. Within an epoch, minibatch after minibatch. Within a minibatch, backpropagation on each training sample. Within backpropagation, the forward pass and the backward pass. Within each pass, layer by layer, neuron by neuron, calculations. Stochastic gradient descent (SGD) is the umbrella term for this algorithm. It does wonders eventually; AI finally learns how to distinguish cats from dogs, to wake up when you say "Hey Google," and to translate from English to German.

> *AI models have to navigate a very high-dimensional, fluid optimization landscape to find near-optimal values for its learnable parameters. The SGD algorithm consists of taking small gradient steps toward the valleys of this landscape.*

2.3.1 Handling Difficult Landscapes

SGD is wonderful but not the final word. Steps can be noisy and the AI model may zigzag too much in its steps. Furthermore, having the same learning rate α for all the parameters can be problematic, especially in narrow valleys such as Yosemite. Instead of wasting time by going back and forth between the opposing cliff walls, one needs to walk down the valley, which is flat but at times slopes down almost imperceptibly. The two directions should not be treated in the same way.

To handle the first problem of noise, we smooth out the descent by keeping a history of gradients. For each parameter, the past gradients are added up with a decay term $\gamma < 1$ which gives less weight to gradients further in the past. To see this, consider the recurrence:

$$a_t = \gamma a_{t-1} + b_t,$$

which unfolds to the exponentially decaying average

$$a_t = b_t + \gamma b_{t-1} + \gamma^2 b_{t-2} + \ldots + \gamma^t b_0.$$

This idea is used in momentum-based methods, where an exponentially decaying average is added to the current gradient:

$$g_t = \gamma g_{t-1} + \alpha \frac{\partial \mathcal{L}}{\partial w},$$

$$w_t = w_{t-1} - g_t,$$

where t is the time step. Nesterov's momentum method is a variation on the vanilla momentum idea in which the gradient is computed at the predicted future position rather than at the present point.

To handle the second problem of adapting the learning rate to each parameter, statistics of past gradients for each parameter are kept. If a parameter has been fluctuating too wildly in the past, we reduce its learning rate to discourage moving in that direction because it is wasteful to move in that direction. On the other hand, if the parameter hasn't changed much in the past and it changes unexpectedly now, then it carries more information, for which we increase its learning rate. The square root of the exponentially weighted average with decay factor $\gamma < 1$ of the squared past gradient values for a parameter w is one way to measure its fluctuation:

$$\sqrt{g_{w,t}^2 + \gamma g_{w,t-1}^2 + \gamma^2 g_{w,t-2}^2 + \cdots + \gamma^t g_{w,0}^2}.$$

The learning rate for w will inversely depend on the above. Methods which compute an adaptive learning rate for each parameter come in several flavors such as AdaGrad, AdaDelta, RMSProp, and Adam; see [2, 62]. These methods often lead to faster convergence compared to the vanilla SGD. The jury is still out to decide which one is the best. Due to the wide variety of optimization landscapes, it is likely that each method will find some landscapes where they will excel.

2.3.2 Stabilization of Training

It does matter where SGD starts from. If we start far away with very large parameter values, then the neuron activation (output) values increase exponentially with the depth in the forward pass, leading to exploding gradients in the backward pass. We don't want to be up in the Himalayas with steep cliffs with large gradients. If we start with very small parameter values, the opposite situation occurs. The gradients are too small and they vanish. We find ourselves in an area which is flat. We don't want to be on Salar de Uyuni salt flat with almost zero gradient.

The trick is to find a balance between the two extremes. Gradients should be stable across neurons and across layers. Since gradients depend on the activation values of neurons, we can stabilize the gradients by stabilizing the activation values. We strive for the zero mean and unit variance distribution for activation values for each neuron.

It is routine to normalize the input features in classical ML to zero mean and unit variance. The same idea can be applied to neuron activation values, which are intermediate hierarchical features. If a neuron has n incoming neurons, then if the weight on each incoming edge is initialized randomly with a normal distribution with zero mean and variance $1/n$, then it will meet our goal of standardizing the activation values. This is *Xavier initialization*. For ReLU, assuming that half of the input neurons are activated, the variance is doubled to $2/n$.

As SGD progresses, the optimization landscape changes from minibatch to minibatch. The underlying reason is that different minibatches have different distributions in the input space. Therefore, the distribution of activation values drifts away from zero mean and unit variance to some nonzero mean and non-unit variance which fluctuate from minibatch to minibatch. If the landscape is very fluid, then SGD struggles.

The *batch normalization* method is used during the training to correct this fluidity of optimization landscape. For each neuron, we compute the mean and the variance of its activation over a given minibatch. By subtracting out the mean and dividing by the standard deviation, we normalize its output:

$$z = \frac{y - \mu}{\sigma}.$$

This turns elongated, stretched contours in the statistical distribution of activations into more circular ones. After this adjustment, we linearly transform the normalized output z, before applying ReLU or the sigmoid activation, by two learnable parameters β and τ (for each neuron):

$$r = \beta z + \tau,$$

intended to modify the output z further in case the data-driven learning finds it useful. The learning process can even determine that the batch normalization is hurting and therefore its effect should be undone by making $\beta = \sigma$ and $\tau = \mu$. Batch normalization is frequently used in practice to stabilize the training. Empirical results suggest that because of batch normalization, the training is robust even if the data distribution were to change from minibatch to minibatch. However, the exact mechanism by which batch normalization helps the training is yet to be understood well. At test time, when we don't have minibatches, we just plug in the mean and the variance computed over the training set.

Xavier initialization nudges SGD to start in a more tractable region of the optimization landscape. Batch normalization can be viewed as the addition of new normalization layers to reduce the fluidity of the optimization landscape. Both of these techniques are primarily empirical, statistical methods; see [58, 100]. One day we will learn more about the optimization landscape to either strengthen the arguments for their usage further or to come up with altogether new robust techniques for training.

SGD faces a variety of challenges when it seeks near-optimal minima in the optimization landscape. Variations of the SGD algorithm have been developed to smooth out the gradient descent. Techniques to nudge SGD toward more tractable regions of the landscape in the beginning and to control the stochastic variation of the landscape from minibatch to minibatch have been developed which have been found empirically to be very helpful.

2.4   Chapter Summary

A single neuron is the fundamental unit of an AI model, and its construction is inspired by biology. Given an input, starting from the input layer, neurons in an AI model fire in a cascaded fashion all the way to the output layer. It is a graph-induced computation. The loss function is an add-on function after the output layer. In order to train a neural network, the gradient of the loss function has to be computed with respect to the learnable (trainable) parameters of the neural network, which are the weights and the bias terms. The backpropagation algorithm computes this gradient efficiently making use of the graph structure and is divided into a forward pass and a backward pass. SGD uses backpropagation for each minibatch of training samples in training epochs. The learnable parameters are tweaked using the gradient descent steps in a high-dimensional, non-convex optimization landscape, controlled by the learning rate hyperparameter. There are several variations of the SGD algorithm targeted at improving the training. Care should be taken in initializing the learnable parameters well so that neuron activations remain in control and gradients don't explode or vanish. Since the optimization landscape is fluid because it changes from minibatch to minibatch, techniques to normalize the neural activation values during training have been found helpful.

In this chapter, we developed intuition into how SGD works. At this point, it is not obvious why it works, that is, why this training process is able to find a good solution without getting stuck in a local minimum. To dive deeper into this challenging topic, we look at the nature of optimization landscapes in greater detail in Chap. 4. In the same chapter, we will learn how exactly the cascaded firings of neurons in a ReLU neuron is able to carve out complex manifolds.

Problem Set 2

1. Why does ReLU lead to faster training compared with the conventional activation functions (the sigmoid and the hyperbolic tangent)?
2. Suppose we are building a fully connected layer which connects n_1 neurons to n_2 neurons. How many learnable parameters does this fully connected layer add to the total number of parameters?

3. What are the partial derivatives of the monomial $f(x, y) = x^n y^m$ with respect to x and with respect to y. Differentiate $(\sin x)^2$ using the chain rule.
4. Suppose a neuron has n incoming neurons. Why does initializing the weight on each incoming edge to zero mean and variance $1/n$ normalize the activation value? Apply the rules from statistics for the variance of the sum and of the product of independent random variables.
5. When does the normalization of the input by subtracting out the mean of pixel values and dividing by their standard deviation help?
6. Suppose you increase the size of minibatches by a factor of four. Should you modify the learning rate? How?
7. Consider the following computation steps:

$$r = w_1 x,$$

$$s = w_2 y,$$

$$t = w_3 r + w_4 s,$$

$$p = \text{Sigmoid}(t),$$

$$L = -\log p.$$

Assume the natural logarithm. Suppose $x = 1$, $y = 2$, $w_1 = -1$, $w_2 = 3$, $w_3 = 5$, and $w_4 = 1$. Perform the forward and backward passes showing all the intermediate values and derivatives. Hints: (1) See Exercise 6 to directly compute the derivative of L with respect to t. (2) If you get the value

$$\frac{\partial L}{\partial w_1} = -1.344707$$

then you are on the right path. Show all the steps reaching to this value.

Chapter 3
Images and Sequences

Space and time are the framework
within which the mind is constrained
to construct its experience of reality.

Immanuel Kant

Some aspects of our models are
inspired by neuroscience, but many
components are not at all inspired by
neuroscience, and instead come from
theory, intuition, or empirical
exploration. Our models do not aspire
to be models of the brain, and we don't
make claims of neural relevance. But
at the same time, I'm not afraid to say
that the architecture of convolutional
nets is inspired by some basic
knowledge of the visual cortex.

Yann LeCun

Previous chapters have laid out the foundations of AI for us in terms of the
overall training process. We have an insight into the expressive power of deep neural
networks, for which we used FFN with several hidden layers of neurons which are
fully connected. Before the dawn of AI, shallow FFNs with one hidden layer were
used in the feature space by ML practitioners. However they don't work well for
images and sequences.

For AI tasks in which we want to understand images, speech, and text, deep
neural networks have been customized for these particular domains. For images, the
fundamental atomic unit is a pixel in 2-D space. Images are made of parts which
combine to form bigger parts. For speech, the fundamental unit is a time interval.
For text, it is a character or a word in a sequence of words. In images, we have

© The Author(s), under exclusive license to Springer Nature Switzerland AG 2021
S. Dube, *An Intuitive Exploration of Artificial Intelligence*,
https://doi.org/10.1007/978-3-030-68624-6_3

spatial dimensions. In speech, we have the temporal dimension. In text, we have a sequence of symbols which can be unbounded with long-term dependencies.

Over time, researchers have figured out how neural networks can be designed to process spatial, temporal, and sequential data. In this chapter, we develop intuition into the customization of AI models, which has proved to be critical for successful applications in different domains.

3.1 Convolutional Neural Networks

The universe has a strong hierarchical design of objects in which coherent units combine to form more complex structures. Objects have parts which are made of smaller parts. We provide two motivations for Convolutional Neural Networks (CNNs), the dominant model for computer vision applications, which is based on a part-whole hierarchy.

3.1.1 The Biology of the Visual Cortex

The first motivation behind CNN is biological. Let us visit Harvard University of the 1950s and 1960s, when David Hubel and Torsten Wiesel described neuroscience of vision in mammals; see [97]. Based on experiments on cats, they classified neurons in two categories:

1. Simple cells which respond to a specific image feature of an oriented line in a particular spatial location.
2. Complex cells which respond to the same oriented line in a spatially invariant manner, that is, they respond to the line irrespective of the spatial location of the line.

Complex cells provide higher-level processing which builds on top of the output of simple neurons, and makes the sensory processing translation invariant. Hubel and Wiesel were awarded Nobel Prize of physiology and medicine in 1981 because of their groundbreaking work.

---

* Exercise 7   Hubel-Wiesel

Based on the research by Hubel and Wiesel on the mammalian visual cortex, discuss how it can suggest a design of neural networks for vision problems. Their breakthrough was serendipitous. Do you know why?

3.1.2   Pattern Matching

The second motivation for Convolutional Neural Network (CNN) comes from signal and image processing, where convolution is a multiply and add operation in a sliding window manner. In linear algebra, the multiply and add operation is referred to as dot product or inner product. The dot product measures the strength of a match between two vectors, and normalizing it with the magnitudes of the vectors yields a correlation coefficient in statistics also known as the cosine distance. Convolving a signal with a filter is one of the very first topics which is taught in introductory signal processing classes. In signal processing, convolution includes an additional step of flipping a filter prior to the operation. In image processing, template-based methods based on the convolution operation for matching spatial patterns have been used to find occurrences of a specific pattern in an image.

See Fig. 3.1 for an example of image processing using convolution. Suppose the image is $H \times W$ and the convolutional filter is $w \times w$. Let B be the border padding to take care of convolution at the border of the image. Let the sliding window stride by which the filter is spatially shifted be S. Then the width and the height of the output will be, respectively:

$$(W - w + 2B)/S + 1,$$

$$(H - w + 2B)/S + 1.$$

In Fig. 3.1, $W = H = 7$, $w = 3$, $S = 1$, and $B = 0$. Therefore, we get 5×5 output.

Commonly occurring local patterns in small patches of images can be detected through the process of convolution. In Fig. 3.1, vertical edge patterns are being detected. A question arises: can there be a finite set of convolutional filters which will detect most of local patterns seen in most of digital images? See the exercise below for an answer.

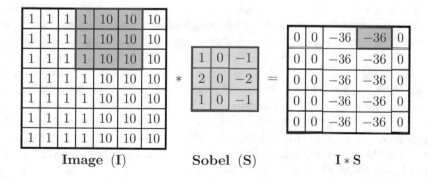

Image (I)   Sobel (S)   I * S

Fig. 3.1 2-D convolution of an image with the Sobel vertical edge detector filter. The convolution operation has been used widely in signal and image processing

** Exercise 8   Olshausen-Field

In 1996, Bruno Olshausen and David Field published a paper in the journal *Nature* titled "Emergence of simple-cell receptive field properties by learning a sparse code for natural images." What does the paper show regarding simple cells in the visual cortex which respond to specific image patterns? See [154].

Therefore, a convolutional layer which learns local image features, such as edges, textures, and color blobs, is analogous to simple cells. Instead of being fully connected, the layer is locally connected, with significant reduction in the number of parameters, which are kept unchanged during the convolution operation across the entire image. The next convolutional layer, which pools the outputs of the first layer, is analogous to complex cells.

3.1.3   3-D Convolution

One can continue the process of increasing the complexity of visual cells further to build higher-level complex cells which respond to more complex patterns. The receptive field of a neuron is the area of the image it is looking at, and it gets bigger as neurons pool outputs of lower-level neurons with smaller receptive fields. This is the core idea behind CNN. A convolutional layer applies the convolution operation on the output of the previous layer. The size of the convolutional filter is the receptive field of neurons in the first layer. The activations of adjacent neurons are pooled by the pooling layer. The next convolutional filter works on the output of the pooling layer. Its receptive field will be the union of receptive fields of all the neurons in the first layer it is dependent on. Activation ReLU is applied after the convolution operation, which lets only positive values pass to the next layer.

Note that unlike 2-D convolution in Fig. 3.1, we work with 3-D convolutions in CNN as each input is a $W \times H \times C$ blob (tensor) with volume, where W and H are the spatial dimensions and C is the number of channels. A channel is referred to as a feature map. An RGB image is a blob of dimensions $W \times H \times 3$ because there are three color channels. Convolving it with a 3-D filter gives $W' \times H' \times 1$ channel (a single feature map). K such filters will produce a $W' \times H' \times K$ blob. See Fig. 3.2. A max pooling layer is a nonoverlapping sliding window operation which takes the maximum value over each region. Note that we can shrink (downsample) spatial dimensions of the output blob by a factor of 2 by applying a 2×2 max pooling, or by using a stride of 2 during the sliding window operation. In a CNN, there are several convolution layers; see Fig. 3.3 which shows the first two layers, intermixed with the downsampling operations. Though it is common to see the spatial dimensions

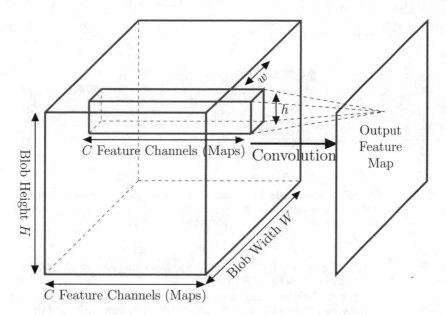

Fig. 3.2 3-D convolution has the same sliding window operation as 2-D convolution (see Fig. 3.1). The position of a $w \times h \times C$ convolutional filter is shown in blue. One performs pointwise multiplication of all the elements of the blob inside the blue cube with the filter coefficients, and then adds them all. The filter slides over the whole image. Therefore, the illustrated red 3-D $W \times H \times C$ feature blob is converted into a flat 2-D $W' \times H'$ feature map. Typically, this is repeated with K different filters, yielding a new 3-D $W' \times H' \times K$ feature blob. A convolutional layer is a blob converter

of blobs typically get reduced through the use of max pooling, it is by no means required. At the cost of extra computation, it is totally fine to preserve the spatial dimensions.

Therefore, the architecture of a CNN can be written as a composition of these layers, for example:

$$y = F(x) = S(FC8(FC7(FC6(C5(C4(C3(M(C2(M(C1(x))))))))))),$$

where $C1$ through $C5$ are five convolutional layers, $FC6$ through $FC8$ are three fully connected layers, M is a 2×2 max pooling layer, and S is a softmax classification layer. In a code, a simple convolutional neural network may look similar to the following:

```
model = SequentialNeuralNetwork(
    [
        Input(shape=input_shape),
        layers.Convolution2D(64, kernel_size=(3, 3),
        activation="ReLU"),
        layers.MaxPooling2D(pool_size=(2, 2)),
```

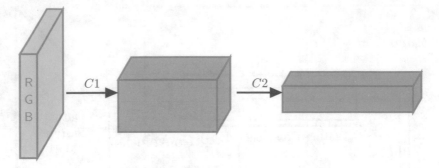

Fig. 3.3 A CNN can be viewed as a device to transform a feature blob (tensor) into another. Spatial dimensions of blobs typically get reduced through the use of max pooling or a stride factor. The depth of blobs can increase or decrease as determined by the number of filters

```
layers.Convolution2D(128, kernel_size=(5, 5),
activation="ReLU"),
layers.MaxPooling2D(pool_size=(2, 2)),
layers.Flatten(),
layers.Dropout(0.2),
layers.Dense(num_classes, activation="softmax"),
]
)
```

In this book, in the context of describing an overall solution, we will be compactly writing $F(x)$, where F will be assumed to be a deep neural network.

For a convolution filter, weights are shared across all the sliding window positions, that is, it is the same filter. To enforce that they stay equal, we update them by the same amount, which is the sum of the gradients during the gradient descent steps of SGD. Note that the convolution layer provides translation-invariant learning of features. Max pooling provides invariance to minor jiggling at every sliding position. This is the reason why a combination of a convolution filter and a max pooling layer has worked very well in practice. Note that for the purpose of batch normalization of a CNN layer, all the neurons in a feature channel are considered the same because they apply the same convolutional filter, and therefore we normalize activation values across the samples in the minibatch as well as the spatial locations. If we just consider one single sample and normalize just across spatial locations for each channel, it is called instance normalization or contrast normalization. One can also normalize across groups of feature channels, including the entire feature blob of the layer.

> A CNN learns to build a representation of an image using a part-whole hierarchy and translation invariance. They are inspired by the biology of visual perception. They can be viewed as a blob (tensor) converter in which neurons learn features of regions of images corresponding to their receptive fields.

3.2 Recurrent Neural Networks

Images are of fixed size, but sequences are of variable length and can run forever, and there may be long-term dependencies. We can't use CNN unless we make the sequences of the same size, truncate them, or segment them into chunks. Another alternative is to use causal convolutions, which is the same as using bounded context from the past. Let's develop intuition into Recurrent Neural Networks (RNNs) which are designed to handle sequential data.

3.2.1 Neurons with States

Let's say the first multidimensional input x_1 in a data sequence arrives and a CNN or an FFN is applied, and at the same time, each hidden-layer neuron computes a value, which is saved. Call this value the "state" of the neuron. At the second step, the next input item x_2 arrives and the same neural network is applied. The stored previous state becomes an input to the hidden neurons in addition to x_2. A new state is computed, which is saved for the third time step. Note at the first time step, there is no previous state and we can let it be zero. See Fig. 3.4.

Therefore, we have the following recurrence:

$$h_t = f(x_t, h_{t-1}), h_0 = 0,$$

$$y_t = g(h_t),$$

where f and g are learned functions implemented by DNN layers and all variables are vectors. In order to train an RNN, we can unroll the recurrence through time. For example:

$$y_2 = g(h_2) = g(f(x_2, h_1)) = g(f(x_2, f(x_1, h_0))).$$

Unrolling yields a directed acyclic computation graph. In deep Recurrent Neural Network (RNN), there are several hidden layers and we have a multilayer architecture:

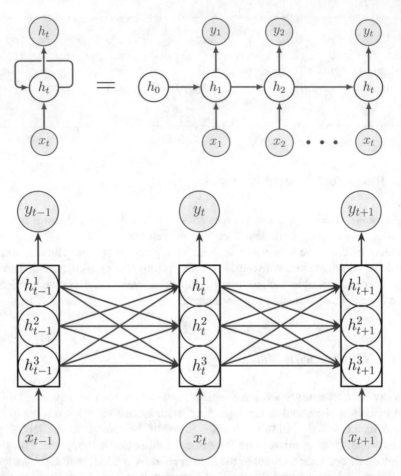

Fig. 3.4 A recurrent neural network (RNN) and its unfolding in time. At the top, all arrows are implemented by multilayer learned functions. For example, $y_t = g(h_t)$ defines a mapping from the hidden layer to the output, where the function $g(h_t)$ is implemented by DNN layers. At the bottom, we zoom in on the red arrow, which is implemented by the shown connections and is a part of the function $h_t = f(x_t, h_{t-1})$

$$h_t^1 = f_1\left(x_t, h_{t-1}^1\right),$$

$$h_t^2 = f_2\left(h_t^1, h_{t-1}^2\right),$$

$$y_t = f_3\left(h_t^2\right).$$

One can construct large, highly complex, multilayer, deep architectures as long as causality of computation is preserved, that is, the computation graph remains an acyclic graph. Weights are shared across time. To enforce that they are equal, we update them by the same amount, which is the sum of the gradients. This is backpropagation through time with weight sharing. The first hidden state h_0 need not be initialized to zero, and we can run backpropagation all the way to h_0 as if it is a learnable layer.

3.2.2 The Power of Recurrence

RNNs are very powerful because of their memory mechanism, and they can simulate a Turing machine making them computationally universal. They are significantly more powerful than Hidden Markov Models (HMMs), a classical tool for sequences. Exactly one state of an HMM is active at a time, whereas for an RNN, all states can be active. Consider the binary case of activation, in which a neuron can be in either of two states, active or inactive. If there are N states in an Hidden Markov Model (HMM), then it has memory of size $\log N$ because there are N unique state patterns. For an RNN with N hidden neurons, the memory size is 2^N. RNNs have feedback loops and are dynamical systems; therefore they can have attractors and chaotic dynamics. This causes sensitivity to the initial conditions, and training can proceed in different directions depending on the initialization. If they get unrolled back in time by a large amount, they can suffer from the vanishing gradient problem because the depth of the computation graphs becomes very large.

3.2.3 Going Both Ways

RNNs are generalized to bidirectional version when the full temporal context is available. For each direction, one employs a unidirectional RNN. Therefore, there are two independent unidirectional RNNs, f_1 and f_2, one working in the forward direction (from the left to the right) on the input sequence:

$$x_1, x_2, \ldots, x_n,$$

and the other working in the opposite direction (from the right to the left).

The outputs of the two RNNs are concatenated together:

$$(f_1(x_1), f_2(x_1)), (f_1(x_2), f_2(x_2)), \ldots, (f_1(x_n), f_2(x_n)),$$

and then input to higher layers of the RNN, which produce the output. The reader may ask if it is fine to use a CNN when full context is available. The answer is affirmative, as we will see later in natural-language understanding (NLU) and speech applications. Time is just like a spatial dimension.

3.2.4   Attention

When an RNN is producing its output, it keeps track of the past implicitly in its
hidden state. What about explicitly saving the past? Suppose we keep the past
states in a memory bank called attention memory, and a mechanism to access this
memory selectively is built. This is called the attention mechanism. The intention
is to include certain relevant states from the past during the computation at the
present time step. The attended memory becomes an additional input to the RNN at
every time step. This will become practically infeasible eventually when the past is
unbounded. Therefore, in practice, it is employed when there is a bound on the
context. The idea of attention has found use in an architecture called encoder-
decoder, which provides a powerful abstraction for many sequence-to-sequence
tasks. See [34, 206] to learn how they have been effectively employed in machine
translation.

   See Fig. 3.5. First consider it without the optional attention memory. There is
an encoder, which produces an internal representation of its bounded input. In the
figure, it receives a sequential input at three time steps, which is processed by a
DNN, often a unidirectional or a bidirectional RNN, to produce an encoding R of
the whole input. For example, R could be the last hidden state of a unidirectional
RNN. Alternatively, R could be the output of a learned function of the hidden states
of a bidirectional RNN. The decoder is an RNN and takes R as the initialization of

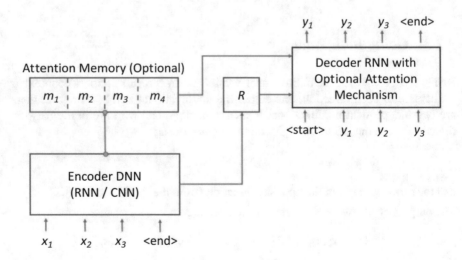

Fig. 3.5 An attention mechanism is an optional add-on to a general encoder-decoder architecture.
It involves saving states of the encoder in a memory to be selectively used later by the decoder.
There are several variations on this basic idea. The figure shows a general view of the approach.
Note that the encoder DNN could be an RNN or a CNN. The encoder produces a representation
R of the input. Time and again, we will see the above architecture employed for applications in
several domains such as NLU, speech, and vision

its hidden state h_0; it produces its first output on the start signal. Note that at every step of decoding, we feed the previous output as the input to the next decoding stage.

Consider the same architecture with the attention memory and suppose the memory bank has N vectors. Consider time step t at the decoding. A method which works well in practice is to implement attention by taking the previous hidden state h_{t-1} of the decoder RNN, computing its attention score with each of N memory vectors in the attention memory using the dot product. A weighted combination c_t of all N memory vectors is computed where the weights are the softmaxed attention scores. The weighted combination c_t is called the context vector. There can be several choices for the score function to compute the attention weights. The score function can be learned by a neural network, or it could be a predefined function such as the dot product. The context vector c_t is appended to the input at the current decoding stage. Here are the steps:

$$h_0 = R,$$

$$\alpha_i' = f(h_{t-1}, m_i), i \in \{1, 2, \ldots, N\},$$

$$[\alpha_1, \alpha_2, \ldots, \alpha_N] = \text{softmax}(\alpha_1', \alpha_2', \ldots, \alpha_N'),$$

$$c_t = \sum_{i=1}^{N} \alpha_i m_i,$$

$$y_t = \text{Decoder}(y_{t-1}, c_t),$$

where m_i is the i-th memory cell in the attention memory, f is the scoring function, and α_i is the attention weight. We can compactly express the attention mechanism as a function A:

$$c_t = A([m_1, \ldots, m_N], h_{t-1}).$$

Therefore, at the highest level, we have the following compact high-level expression of the underlying architecture:

$$(R, M) = F(x),$$

$$y = G(R, M),$$

where M is the attention memory, F is the encoder, and G is the decoder, and the attention function A is an internal function of G.

Though the attention mechanism may look complicated at first sight, it is really a modification of the basic multiplication and add operation in AI:

$$y = b + \sum w_i x_i,$$

where now the weights are dynamically computed using the data at the runtime, $w_i = f(x)$.

> The encoder-decoder architecture has proven to be successful across different application areas of AI. Some examples are language translation, image captioning, activity recognition, object detection, speech recognition, anchored speech recognition, speech synthesis, and handwriting recognition.

There are several variations on this theme, and the opportunities to modify the architecture are unlimited. Can we use h_t from a lower layer in a multilayer decoder RNN as the input to the scoring function? Is it necessary to use an RNN as the encoder? Can we input R to every stage of the decoder? Can we use a CNN for the encoder? Can there be other forms of attention? There is no limit on one's creativity here. Time is just a dimension like space and one can creatively use a CNN. The output is a feature blob in which the spatial position takes the place of time. We pay more attention to certain spatial positions than others. In other words, if the feature blob is of spatial dimensions $W \times H$ and has C channels, then the attention mechanism A is given by

$$c_t = A([m_1, \ldots, m_N], h_{t-1}),$$

where $N = W \times H$ is the total number of spatial positions. For each spatial position, the C-dimensional feature fiber is an attention memory vector; see Fig. 3.6.

The input could be an image such as a spectrogram for Automatic Speech Recognition (ASR). The input could be a photograph and the goal of the decoder to generate an image caption. Here we have a mix of computer vision and the

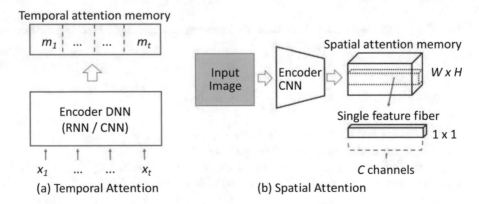

(a) Temporal Attention   (b) Spatial Attention

Fig. 3.6 Time and space are qualitatively really the same. Therefore, the temporal attention mechanism generalizes to spatial attention. Feature fibers in a feature tensor which is output by a CNN are the spatial attention memory vectors. The attention mechanism selectively retrieves different parts, either in time or in space, of the encoded input

NLU task of automatically inferring a caption of a given image, where the encoder is a CNN and the decoder is an RNN combined with an attention mechanism. This provides a nice way to merge the two fields of computer vision and Natural Language Understanding (NLU); see [247]. For pure NLU tasks, the input could be the matrix of embedding vectors viewed as an image which is fed to an encoder CNN.

3.3 Self-Attention

The above attention mechanism allows the decoder to refer back to the computation done by the encoder; therefore it is encoder-decoder attention. The encoder can also pay attention to its own previous hidden states while it is performing the encoding. This will work if the encoder works on a finite chunk of data or if a finitely many previous states are saved in the memory. This will constitute encoder-encoder attention. On a bounded chunk of input, the attention mechanism is generalized to self-attention by computing all pairwise, bidirectional, order-aware attention scores; see [218] for an introduction to self-attention. In a similar way, one can define decoder-decoder attention, which has to be causal and unidirectional by definition. Recent NLU techniques employ all three forms of attention; see [218].

An analogy is communication between two people, A and B. Person A (encoder) is given a report and has to summarize it for B, and B has to immediately verbally respond to it. Person A gives self-attention (encoder-encoder) while preparing the report, makes notes on the margin of the report, and produces a summary, which is an encoded representation R of the report. Person B (decoder) takes the summary R and produces an on-the-fly verbal response, paying attention to the margin notes made by A (encoder-decoder attention). The margin notes are the attention memory in this analogy. During the verbal response, B is paying attention to his/her own train of thoughts and previously uttered words (decoder-decoder causal attention) before speaking the next word.

Self-attention is the basis of a new architecture of deep networks which is known as a transformer. A transformer can be viewed as a generalization of a feed-forward network which processes input tokens in parallel using self-attention layers. The processing is order-aware because the positions of the words are also encoded in their representations. See [48, 218] and Sect. 10.5 to learn how self-attention can be used in NLU. See Sect. 9.4 for its application in computer vision.

3.4 LSTM

To solve sequence problems more effectively with stable training, the hidden state of a neuron is generalized to a memory cell. The values can be read from and written into the memory register, and its contents can be retained or forgotten. These operations have to be differentiable for the backpropagation algorithm to work. Thus the controlling gate values are in the range [0, 1] and the operations are soft. At the extreme values, they are like "on" or "off." The update of a memory cell is

$$\text{Contents} \leftarrow W \times (\text{Proposed Contents}) + F \times (\text{Old Contents}),$$

where W and F are the *write* and the *forget* gates, respectively. Long Short-Term Memory (LSTM) is such an RNN and a Gated Recurrent Unit (GRU) is another variant. Both LSTM and GRUs work better than simple RNN and are able to capture long-term dependencies with short-term memory cells.

In this book, it is assumed that an RNN cell is either an LSTM cell or a GRU cell, and the term RNN subsumes all different forms. Therefore, whenever we say RNN, the choices for its implementation will include LSTM and a GRU. Simple RNN without gated memory cells can be used for simple tasks with short-term dependencies.

Imagine you are talking to someone important to you. You think the person is happy with you based on the conversation. But in the end, a comment is made "I am so disappointed with you." Well, you will forget what was said earlier. In the same way, your spouse says to you "I am so happy today," and then it is followed by a recounting of some difficulties he/she is facing. Despite the complaints, overall you will continue to rate the tone as positive because you are anticipating something positive, and you have decided to retain that first statement "I am so happy today" in your memory. LSTM captures this information flow across time which gets pulled in at every time step for computation, and this flow can be terminated anytime and replaced by new information.

An idea to control the information flow in a neural network which is inspired by LSTM is that of gated neurons, in which the final output y' of a neuron implementing the function $y = F(x)$ is controlled by a learned sigmoid gate G as follows:

$$y' = Gy + (1 - G)x.$$

If $G = 0$, it becomes like a pass-through neuron. A network using this mechanism is called a highway network; see [202].

The idea of keeping memory is essential for building intelligent systems, and one can generalize the idea of local memory in an LSTM cell to a global memory which retains facts. If one is building a question answer (QA) system in NLU, then it is natural to adopt this approach so that one can refer to an appropriate portion of the input text in order to retrieve the right answer; see [241] for this idea implemented in the form of memory networks. Having a conversational memory is also important in chatbots.

An LSTM-based RNN learns long-term dependencies in sequential data. When combined with an attention or self-attention mechanism, it provides a powerful method to solve problems involving sequential data.

3.5 Beyond Images and Sequences

CNNs were used successfully in [124] for MNIST handwritten-digit recognition, and RNNs were introduced in [182] and LSTM in [93]. All have been highly influential in AI and have worked out admirably in building successful applications. In less than two decades after the 1998 LeNet paper, we have had great success in applying CNN to computer vision problems. See [116] for the first major breakthrough of AI in computer vision, and also see [79, 211] for other noteworthy work. For attention-based RNN models, see [9, 137], and for self-attention, see [218]. We will be studying these in more detail in later chapters on computer vision and NLU.

Let's now think beyond images and sequences of words. Here are a few scenarios:

Videos   Image sequences have their own particular characteristics along the dimension of time which make them more than just 3-D image data. The goal is to perform computer vision tasks on videos, an example of which is activity recognition.

3-D Point Cloud Data   Self-driving cars and robots depend on depth sensors, which produce 3-D point clouds. One can obtain range images, which are 2-D images showing the distance or depth of each point in a scene. The goal is to use the depth data, in the form of either 3-D points or range images, to solve computer vision tasks such as object detection and scene segmentation.

3-D Volumetric Data   Medical images such as CT scans produce volumetric data. The goal is to detect and segment out areas of interest.

Manifolds   Consider climate data for the surface of the Earth, which has a 3-D spherical structure. The goal is to extend convolution to smooth manifolds so that one can make weather predictions, for example, for cyclones. Another example is a 3-D object with a smooth surface, and the goal is to learn pose-invariant features for shape matching and retrieval.

Graphs   A graph consists of nodes (vertices) and edges. Each node has a feature vector associated with it. The goal is to produce a hierarchical representation of the graph and compute global features of the entire graph or to classify subgraphs.

Data Structures   In the most general case, one can have complex data structures, and the desired output is based on nontrivial algorithms. These could be lists, trees, or graphs. The goal is to perform complex reasoning on the provided data structures.

Symbolic Mathematics   Consider problems in symbolic mathematics. The goal
is to use AI to solve differential equations and to perform integration. The input
is an equation or a mathematical expression.

How can we generalize the AI models for these diverse cases?

We will discuss the use cases of videos, 3-D point cloud data, and 3-D volumetric
data in Chap. 9.

Graphs are ubiquitous in the sciences and engineering, and CNNs have been
generalized to graphs; see [46]. Consider a graph in which nodes have feature
vectors. A *graph CNN* updates these by convolution filters applied on neighboring
nodes. That is, the receptive field of the convolution filter is a suitably defined
neighborhood of each node. We can write this as

$$y = F(N(v)),$$

where $N(v)$ is the neighborhood of a graph vertex v, and F is a learned graph
convolutional filter which takes the features of vertices in the neighborhood and
produces a new feature vector for v; see Fig. 3.7. One can apply graph convolutional
networks on 3-D point cloud data, where there is a natural concept of neighborhood
of points. It is possible to be creative for other applications where there is
unstructured data with a concept of local relationships. For example, one can employ
graph CNNs to predict the behavior of multiple agents such as cars and pedestrians
in self-driving car technology. The standard convolutional network for images is a
special case of graph CNNs in which we have uniform grid-like neighborhoods.

For *convolutions on surfaces* such as the Earth's surface or 3-D shapes, the idea
is to have a local geodesic system of polar coordinates to extract patches around
a point, very much like a placing a small spider web around it; see [140]. Again,
there is a concept of locality as defined by the surface of a 3-D object such as
the Earth, and one applies flexible convolutional operations. This framework can
be generalized to arbitrary manifolds; see [40], which employs ideas from modern
physics.

For general data structures and algorithms, the Turing machine is a computationally universal model. We can have a neural network motivated by the Turing
machine in which there is external memory; see [66, 69] and Fig. 3.7. The read
and write operations are differentiable soft operations. Therefore when we write,
we write on all memory cells controlled by write gates whose values are in the
range [0, 1], and it is the same for reading. Neural Turing machines (NTMs) and
differentiable neural computers (DNCs) are inspired by these ideas. They consist
of a memory, which is equivalent to the tape of a Turing machine, and a controller,
which is based on AI. They are designed to perform *complex reasoning on general
data structures*. Here are a few illustrative examples:

1. A list of unsorted symbols where each symbol has an attached real-valued
   priority is the input. An AI model is trained to output the sorted sequence.
2. Training data consists of randomly generated London underground maps which
   have L lines serving S stations. The input is in the form of triples such as (Station,

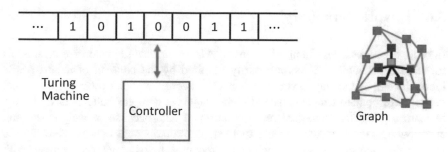

Fig. 3.7 AI models have been creatively generalized to different types of data and to different architectures. Inspired by Turing machines, shown on the left, it is possible to train AI models consisting of a controller, read/write head, and external memory, which can learn more complex reasoning on general data structures. Graph convolutional neural networks work on the ubiquitous data structure of graphs, shown on the right. A local neighborhood of the green node, consisting of neighboring red nodes, is shown

S1; Station, S2; Line, L1). An AI model is trained to answer questions about the shortest path between two stations on such example graphs. Then it is given the actual London metro map, which has $L = 11$ lines serving $S = 270$ stations. A question is asked by specifying two stations (Moorgate, South Kensington, —), and the output is the shortest path between the two specified stations.
3. Family trees. An AI model is trained to answer questions about relationships between family members such as "Who is John's maternal great aunt?"

We are far from learning complex algorithms using backpropagation, but it is wonderful to see that the first steps toward data-driven methods which learn algorithms to solve nontrivial problems using symbolic reasoning have been taken.

An equation is a string of symbols with a structure. Consider a differential equation $xy' - y + x = 0$. Its solution is

$$y = x \log c - x \log x.$$

Given a function $f(x) = x \cos x$, its integral is

$$F(x) = x \sin x + \cos x + C.$$

Since mathematical expressions can be parsed as trees, which can be converted into sequences in prefix notation, therefore seq2seq AI models can be used to solve univariate mathematical problems with remarkable accuracy. *Symbolic manipulation of expressions* can be viewed as a special kind of language translation. See [120] for work in this exciting new direction.

For another novel application of AI, see [216], which derives well-known *equations in physics* by fitting a deep feed-forward network to data from an unknown physics function and checking for well-known symmetries in order to derive the equation. Here an AI model is being used as a function approximator in order to discover symmetries in the function.

3.6   Chapter Summary

In order to build successful applications of AI, it is critical to customize AI models for different domains. CNNs are partly inspired by the concept of a part-whole hierarchy and by the way the mammalian visual cortex works. Lower convolutional layers correspond to simple cells and higher layers to complex cells. A CNN is able to learn a translation-invariant representation of data because of weight sharing, and pooling layers make it robust to local deformations. It can be viewed as a feature blob (tensor) converter in which the receptive fields of neurons increase with depth. For sequential data, RNNs are used which memorize the states of hidden neurons. An RNN cell can be generalized to an LSTM cell, which is equipped with a memory register controlled by soft read, write, and forget gates. These gates enable information flow from the past to be retained, forgotten, or modified. LSTM-based multilayer RNN architectures have been effective in capturing long-term dependencies. Coupled with an attention mechanism in an encoder-decoder architecture, they provide solutions to many NLU tasks. The attention mechanism allows the decoder to refer back to the states of the encoder saved in a memory bank. It can be generalized to self-attention, in which the encoder performs bidirectional, pairwise attention, when building a context representation of the input tokens.

Finally, a number of novel architectures have been proposed for different types of data and problems. For general data structures such as lists, trees, and graphs, they are inspired by the Turing machine model. For symbolic mathematics, data is in the form of equations and expressions, which are amenable to sequential AI models. In Chap. 9, we will look at different types of computer vision data.

Problem Set 3

1. How can we ensure that the spatial dimensions of the output feature map obtained by a 2-D convolution are the same as those of the input?
2. Why can the convolution operation be applied for pattern matching (template matching) in classical image processing?
3. For template matching, how can the convolution operation be made more robust to changes in the contrast as measured by the variance of the pixel values or shift in the overall brightness as measured by the mean of the pixel values?
4. The convolution operation is based on the inner product of two vectors. What is the relationship between the inner product and the correlation coefficient in statistics?
5. Suppose you have an input blob of dimensions $256 \times 256 \times 64$, where 64 is the number of channels. How many parameters would a 2×2 convolutional filter and a 1×1 convolutional filter have?
6. Consider an input blob of dimensions $W \times H \times C$. What is the impact of a 2×2 pooling layer on the dimensions of the output blob? What is the impact of a 1×1 convolutional layer with K 1×1 filters?
7. How does the backpropagation algorithm work on a max pooling layer?

8. What is a better way to reduce the spatial dimensions of a feature blob: a convolution operation with a stride of two or a convolution operation with a stride of one followed by a 2×2 max pooling layer?

9. What will be the size of the receptive field of two successive 3×3 convolutions layers? Is it better to use a 5×5 convolution filter or two successive 3×3 convolutions layers?

10. How can the output of a bidirectional LSTM be used by the higher layers of a model?

11. Suppose you are trying to predict the mood of your friend, family member or significant other. You notice that it depends on your spoken words, some of which go way back in the past. What type of AI model will you use to solve the problem? If you notice that only your most recent N spoken words matter, then how will you solve the problem?

12. Can a conventional CNN be viewed as a graph CNN?

Chapter 4
Why AI Works

> Don't only practice your art, but force your way into its secrets.
>
> Ludwig van Beethoven

> It's fine to work on any problem, so long as it generates interesting mathematics along the way – even if you don't solve it at the end of the day.
>
> Andrew Wiles

Children enjoy watching magic shows. With sleight of hand, an accomplished magician can entertain an audience. People exclaim to each other, "Wow! How did she do it?"

One of the comments on AI has been that we don't really understand what's going inside the black box of an AI model. What are all the hidden neurons doing when a CNN is recognizing a cat? How is an AI model able to generalize to unseen examples? Another persistent comment has been that we don't really know what's happening during the training process. What is the nature of this optimization landscape? Why doesn't the training get stuck in a local minimum?

Behind all these questions is a fundamental question which is primarily motivated by burning curiosity. Why does AI work so well in practice? What is the secret behind its magical performance in ML competitions? In this chapter, we raise the curtain a bit. This is not the final answer by any means. A lot has yet to be discovered, proven, and understood, but we have gained greater understanding with time.

We got a glimpse about the working of AI based on previous chapters when we learned that its expressive power grows exponentially with the depth of AI models. It is time to dive deeper into the topic and build a strong intuition into the underlying mechanisms. The more we understand, the more questions we will ask, the more we will innovate, and the more progress is bound to follow.

4.1 Convex Polytopes

As shown in Fig. 1.3, the basic carved unit of AI is a convex polytope (linear region). To understand how the input space gets divided into these pieces, we will start with a single ReLU neuron. See Fig. 4.1 which shows how a ReLU neuron divides up the input space into two regions. Each region (half-space) is an infinite convex polytope. Each of the regions has an activation pattern associated with it. For a single neuron, it is either active or inactive. Suppose the same input (x_1, x_2) is sent to a second ReLU neuron. Now we will have four convex polytopes (unless the cutting hyperplanes are identical). Each convex polytope will have an activation pattern:

$$(a_1, a_2)$$

where $a_i \in \{\text{active, inactive}\}$ is the activation for neuron $i \in \{1, 2\}$.

Here is another example. In Fig. 2.2, there are five input neurons and three hidden neurons. Each hidden neuron divides up the 5-D input space into two 5-D half-

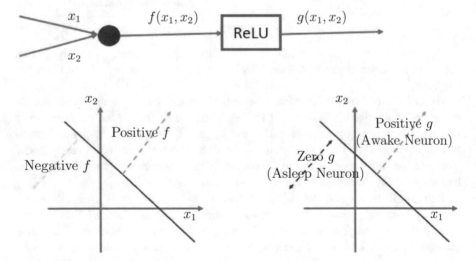

Fig. 4.1 Activation region for a single ReLU neuron in its input space. The ReLU activation function zeros out the negative half-plane of f. We say that the neuron is awake or active in the positive half-plane and it is asleep or inactive in its other half-plane. Functions f and g assume 0 value at the red lines. In d-dimensional space, the lines are hyperplanes. Function f is a linear function and g is non-linear. Function g is piecewise linear

spaces. All three hidden neurons divide up the 5-D input space into many convex polytopes, each polytope having an activation pattern:

$$(a_1, a_2, a_3).$$

In the same figure, there is one output neuron. What does it do to the input space? Though it is clear that it will divide up the 3-D feature space (h_1, h_2, h_3) into two 3-D half-spaces with a hyperplane, it is not obvious what its impact on the 5-D input space is.

4.2 Piecewise Linear Function

Let us unravel the answer to the question of how a neuron subdivides spaces which are carved by neurons in the layers preceding it.

4.2.1 Subdivision of the Input Space

See Fig. 4.2 to develop intuition. In this illustrative example, the input is 2-D. There are two hidden layers h_1 and h_2, each with two neurons, and one output layer O with a single output neuron.

Hidden layer h_1 divides up the input space into four convex polytopes with two lines corresponding to the two neurons. The activation pattern of each polytope is shown in parentheses. For example, the top right polytope has the activation pattern

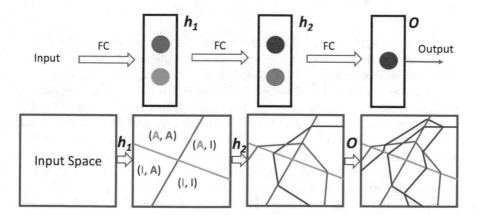

Fig. 4.2 Illustration of how a fully connected DNN carves out the input space. Each layer progressively subdivides the convex polytopes carved out by previous layers using bend-lines. See Sect. 4.2

(A, I), which stands for (active, inactive) pattern. Inside this polytope, the green neuron is active and the blue neuron is inactive.

Consider the purple neuron in hidden layer h_2. Fix the activation pattern of neurons in hidden layer h_1 to (A, I). The purple neuron cuts the convex polytope with activation pattern (A, I) with a line, as shown. It is a linear function as the computation graph of the neural network does not have any non-linearity when you fix the activation pattern. Awake neurons become pass-through neurons, and asleep neurons don't do anything. It is a *composition of linear functions*, which is again linear. Weights get multiplied, and bias terms get added along the active computation paths, which start at the input layer and end at the output.

Move along the purple line from right to left. As soon as we step into the adjacent convex polytope with activation pattern (A, A), the blue neuron wakes up and gets activated. The linear function changes accordingly. It is a new composition of linear functions and the line bends. The change is continuous at the boundary of the two polytopes. We call the bending lines as bend-lines. Similarly, the orange neuron of layer h_2 carves up the input space with another bend-line.

In total, we now have 13 convex polytopes. Each polytope has a 4-D activation pattern:

$$(a_1, b_1, a_2, b_2),$$

where a_1, b_1 are the two neurons in hidden layer h_1 and a_2, b_2 are the two neurons in hidden layer h_2.

The process continues to the output layer O. The red output neuron further carves the input space with a bend-line as it moves from one polytope to another. Follow the red line in Fig. 4.2 as it bends around in the space.

In total, we now have 21 polytopes. Each polytope has a 5-D activation pattern:

$$(a_1, b_1, a_2, b_2, o)$$

where o is the output neuron. Each polytope has a linear function fitted on it. The whole input space has been fitted with a piecewise continuous linear function; see Fig. 4.3 for an illustration.

The above analysis is for AI models which use ReLU as the activation function. This is true for most modern AI models in practice. Sometimes variations of ReLU functions such as ELU (exponential linear unit), SELU (scaled ELU), leaky RELU, PReLU (parametric ReLU), and smooth ReLU (also known as the softplus function $\ln(1 + \exp(z))$) are used. What if the activation is a conventional non-linear function such as the sigmoid function or the hyperbolic tangent function? Then we lose piecewise linearity of the final function. Neurons will have values in the range (0, 1), and there are no on/off activation patterns of neurons. The result is a non-convex, non-linear function. The contour plot will consist of smooth curves. What has been observed is that ReLU-based AI models are much easier to train. It can be safely said that the use of the ReLU function contributed significantly to the success

Fig. 4.3 A piecewise continuous linear function fitted over convex polytopes. The parallel lines show the contour plot of the function for different values. A linear function has parallel hyperplanes as its contours. The gap between the parallel lines will be the same for a fixed step size of contour value

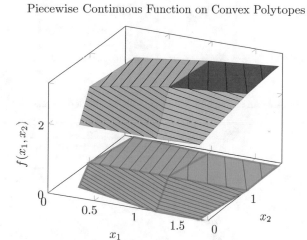

Piecewise Continuous Function on Convex Polytopes

of deep learning, and that's why we have focused on the analysis of ReLU-based models. Does one have to always use ReLU? No, because for certain applications, it is preferable to use a smooth activation function; for example, see Sect. 6.7, where a sine function is used to represent an individual signal.

4.2.2 Piecewise Non-linear Function

In classification, the piecewise linear function will get squished to the range $[0, 1]$. The result is a piecewise non-linear function whose range is $[0, 1]$. The contour plot still has parallel lines, but the function slopes non-linearly over each convex polytope.

Suppose the fitted function gives the probability $P(x)$ that the input x is a cat. The final output is as follows:

$$\text{Decision} = \begin{cases} \text{Cat}, & \text{if } P(x) > T_1 \\ \text{Dog}, & \text{if } P(x) < T_2 \\ \text{Indecision}, & \text{otherwise} \end{cases}$$

Thresholding at T chops off a convex polytope along the contour line with value T. Therefore, each convex polytope will get subdivided into at least one and at most three polytopes, corresponding to cat, dog, and indecision outcomes; see Fig. 1.3.

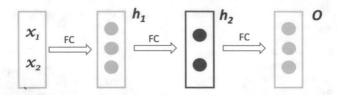

Fig. 4.4 An example FFN with FC (fully connected) layers. See Fig. 4.5 to see how it works

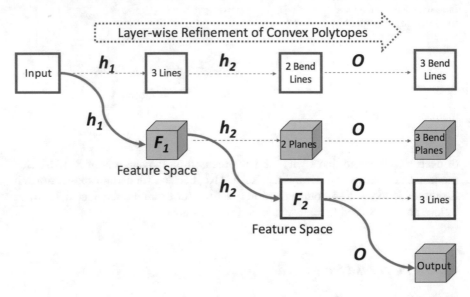

Fig. 4.5 Carving of the input space and the feature spaces of the FFN shown in Fig. 4.4 by subdividing them into convex polytopes using bend-lines and bend-planes. Every layer L is performing a simple carving of the preceding layer $L - 1$, a slightly more complex carving of the layer $L - 2$, and so forth. The solid blue curved arrows show the forward pass. Each row corresponds to a layer. Similarly each column corresponds to a layer. For example, the first row (the input layer) and the fourth column (the output layer O) show how the output layer carves the input space

4.2.3   Carving Out the Feature Spaces

We now build a complete picture. See Figs. 4.4 and 4.5.

In Fig. 4.4 we show a FC-FFN with two inputs. There are two hidden layers h_1 and h_2 with three and two neurons, respectively. The output layer O has three neurons. Each layer is a FC (fully connected) layer. We dive deeper into how each layer divides all the feature spaces (blobs) built by the preceding layers into convex polytopes.

See Fig. 4.5. The 2-D input space is transformed by layer h_1 into a 3-D feature blob F_1, which is again transformed by h_2 into a 2-D blob F_2. Finally, the output layer takes us into a 3-D space. The output of a layer with k neurons is a k-dimensional feature space.

Layer h_1 cuts up the input space with three lines. Layer h_2 divides the feature blob F_1 with two planes and the input space by two bend-lines. Finally, the output layer O divides the feature blob F_2 by three lines, F_1 by three bend-planes, and the input space by three bend-lines.

Every neuron in a layer L is fitting a simple ReLU-activated linear function to the feature space spanned by the neurons in the preceding layer $L - 1$, a slightly more complex piecewise linear function to the feature space spanned by the neurons in the layer $L - 2$, and so forth, till it fits the most complex function to the input layer; see Fig. 4.5. Each neuron in a higher layer is trying to help separate the classes in the preceding feature space by computing a useful feature, which is a "feature of features" and a "pattern of patterns," and therefore it is a higher-level feature. It will be hard for a neuron in the first layer to do so because of complex geometry of manifolds, for which we need to bend around the manifolds.

Suppose we move along a straight-line segment in the input space. What does this linear trajectory map to in the feature spaces computed by the hidden layers? Consider the activation pattern of the hidden layer h_1:

$$(a_1, a_2, a_3)$$

and suppose at the starting point of the line segment, it is

$$(A, A, A),$$

all three neurons activated. When we move in a straight line in the input space, then in the 3-D feature blob F_1, we move in a straight line because affine transformations map lines to lines. Note that the bias terms introduce translation making the transformations affine, a co-linearity preserving superclass of linear transformations. Suppose the activation pattern changes to

$$(A, A, I),$$

with one neuron getting inactivated. Now the trajectory in F_1 bends, and then we move in another straight line in the 2-D plane spanned by the first two dimensions and with the last dimension being zero. If the three dimensions of F_1 are (x, y, z), we are now moving in the (x, y) plane. Suppose the activation pattern changes to

$$(A, I, I),$$

with the second neuron also getting inactivated. We are now moving along the x-axis. Therefore, a straight line in the input space bends around in the feature space.

Straight lines in the input space bend around in the feature spaces.

** Exercise 9   Carving

1. See Fig. 1.3. Specify the architecture of a neural network which could have produced the illustrated carving.
2. The number of convex polytopes (linear regions) in the input space measures the expressive power of an AI model. What could be a method to measure the expressive power in the feature spaces built by the AI model?

4.3 Expressive Power of AI

This section is for readers who want to develop an advanced intuition into why the expressive power of AI models is exponential in the depth of the models. The result is intuitive based on the geometrical carving process we have explained in the previous sections.

The goal is to find the bounds on the maximum realizable expressive power. A result on the upper bound has to show that there does not exist any weight assignment which will make a network exceed that bound. A result on the lower bound has to show that there exists some weight assignment which will make a network reach that expressive power.

We outline a high-level sketch of a proof that the number of convex polytopes (linear regions) is exponential in the depth of an AI model. Suppose the input is d-dimensional. We will keep d fixed. Consider the first hidden layer with n_1 neurons. Each neuron cuts the d-dimensional space with a hyperplane. The total number of convex polytopes $r(n_1, d)$ carved by the first hidden layer of n_1 neurons in the d-dimensional input space is given by a beautiful result from the *theory of hyperplane arrangements*:

$$r(n_1, d) = 1 + n_1 + \binom{n_1}{2} + \binom{n_1}{3} + \ldots + \binom{n_1}{d};$$

see [204] for details. Suppose the second hidden layer has n_2 neurons. Each d-dimensional polytope gets further subdivided by n_2 hyperplanes into $r(n_2, d)$ pieces, which bend around from one polytope to another. Therefore, the total number is given by

$$r(n_1, d) r(n_2, d).$$

Of course, this is the most expressive carving which is possible and which is also desirable. If some bending hyperplanes don't intersect with some polytopes, then the number will be less. Hopefully, over training epochs the carving hyperplanes will shift around to increase the expressive power of the model.

For an example, refer to Fig. 1.3. The first layer has three neurons; therefore we can confirm that

$$r(3, 2) = 7,$$

by looking at the straight dashed lines. The next processing layer, which is the output layer, has one neuron, and it subdivides each of the seven polytopes into two parts with the bending dashed line, since $r(1, 2) = 2$. In total, we have 14 polytopes. The reader should verify that if we had another neuron in this layer, then each of the seven polytopes could have been subdivided maximally into $r(2, 2) = 4$ polytopes, yielding 28 polytopes in total. For another example, see Fig. 4.2. The first hidden layer leads to $r(2, 2) = 4$ polytopes. Each neuron in the second layer divides each of these four polytopes into two parts, but together they achieve maximal subdivision into four parts in only one of the polytopes.

Note that each term is a partial sum of binomial coefficients. There is no known closed formula for the partial sum of the first d binomial coefficients. Note that, for fixed d

$$\binom{n}{d} = \Theta(n^d),$$

and therefore we have

$$r(n, d) = \Theta(n^d),$$

that is, the function grows polynomially in n.

---

* Exercise 10   Theta Notation

What does the theta (Θ) notation used in computer science signify? What are the omega (Ω) and the big O notation for? Show the above stated result, that is,

$$r(n, d) = \Theta(n^d).$$

For two layers, we have the following upper bound on the maximal expressive power:

$$O\left(n_1^d n_2^d\right).$$

Therefore if there is only one hidden layer, then the number of convex polytopes grows polynomially with its width n for a fixed d. If there are L processing layers (including the output layer), we have

$$O\left(n_1^d \dots n_L^d\right),$$

and assuming that all layers have the same number n of neurons, we have the upper bound:

$$O(n^{dL}),$$

which is *exponential* in L. Since n, d, and L are typically large, an AI model has an astronomically high expressive power. This is a remarkable result. Because of its importance, we will call it a mathematical theorem.

Carving Theorem Suppose a feed-forward fully connected L-layer deep network has d-dimensional input and its each layer has n neurons. Then, it carves the input space with $O(n^{dL})$ convex polytopes.

See [171] for the same result. For a detailed formal proof based on tropical geometry, see [253]. Also refer back to Exercise 2.

Is the above upper bound asymptotically tight? Can one choose the parameters of a network in such a way that the exponential expressive power is realized? We can choose the learnable parameters of a network such that multiple input activation regions get mapped to the same region in a feature space computed by a hidden layer. This can be done by $L-1$ consecutive layers to create exponentially many preimages. When the feature space computed by the $(L-1)$-th layer gets carved by the final hidden layer, then the carving gets replicated in all the preimages in the input space. See [145] for this construction and for the lower bound $\Omega((n/d)^{d(L-1)}n^d)$, $n \geq d$, which is exponential in depth and polynomial in width. The division by d is a result of the special hand construction of the model. The first hidden layer folds the d-dimensional input space such that $(n/d)^d$ regions map to the same output region. This folding process is done by the first $L-1$ layers.

4.4 Convolutional Neural Network

We now develop an intuition into how CNNs carve up the input space. See Fig. 4.6. We will focus on the very first convolutional layer.

The same convolutional filter of dimension $n \times n$ is applied in a sliding window fashion across the image. Therefore, the input image has been divided into a set X of $n \times n$ overlapping smaller images. Call X the *sliding window input space*. Convolutional filter operates like a fully connected hidden layer on the sliding window space.

In Fig. 4.6, we show the sliding window input spaces of two images A and B. Each image yields k many sliding windows. Therefore there are k points in this $n \times n$-dimensional space. Suppose some of these are basic edge-like patterns from a visual perspective, some are textured, and the remaining are smooth blob type.

---

* Exercise 11   Sliding Window Space

Given an image and an $n \times n$ convolutional filter, compute the total number of sliding windows, that is, the number k of points in the $n \times n$-dimensional sliding window space.

Suppose we have three $n \times n$ convolutional filters. During training, assume that one filter gets trained to detect edge patterns, the second filter for textured patterns, and the third filter for blobs. See Fig. 4.6, where on the left side, for image A, each point is $d = n \times n$-dimensional and represents an $n \times n$ sliding window region. We have shown points of the three kinds (edge, texture, blob) and the corresponding convolution filter hyperplanes. For image B, these patterns can be in totally different spatial positions in the image. But due to translation invariance, the sliding window space of B will be very similar to that of A; see Fig. 4.6. In the original input space, though these two images may belong to different convex polytopes, the carvings of their respective polytopes are related to each other because of this invariance.

Therefore, the convolutional filters lead to translation invariance in pattern matching.

For higher-level convolutional layers, the argument generalizes. We work with larger sliding window spaces of images, the dimensionality of which is given by

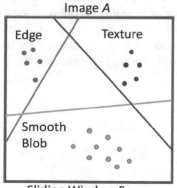
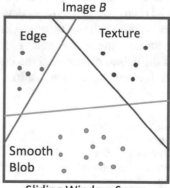

Fig. 4.6 Carving up of the sliding window space by a CNN. Say for an image, an $n \times n$ convolutional filter is applied at k positions. Sliding window space will have k points in a $d = n \times n$-dimensional space

the size of the receptive fields of the neurons. Instead of simple local patterns, convolutional filters now specialize for more complex patterns, which can be viewed as "patterns of patterns." This is exactly how the CNN carves in a compositional manner in order to make use of the part-whole hierarchy in images. For example, a complex pattern may consist of a vertical edge on the left, a texture at the top, and a smooth green blob in the middle. Such higher-level patterns will get clustered in the sliding window space, and a learned convolution filter will detect them. The underlying mechanism shown in Fig. 4.6 remains the same.

We can understand the max pooling layer using the same intuition. Say it is a 2×2 max pooling layer. In the sliding window space, we have four points for a 2×2 pooling window. The point corresponding to the highest function value will be retained, and the other three will be discarded. The highest value means that the point is furthest from the convolutional hyperplane in Fig. 4.6. The points closer to the border will get discarded. Therefore, there is immunity to local perturbations in the sliding window space.

4.5 Recurrent Neural Network

We now develop an intuition into how RNNs carve up the input space.

An RNN can be unfolded in time to turn it into a directed acyclic computation graph. See Fig. 4.7 for an illustrative example. The input at the current time step is (x_3, x_4). At the previous time step, the input was (x_1, x_2). The current output uses the current input and the previous hidden state.

Clearly, this is a special case of FFN with weight sharing. The mechanism of carving up the input space with convex polytopes and fitting them with piecewise linear functions holds fine.

Note that the input space is 4-D. One may ask how the output neuron carves up the 2-D subspace (x_1, x_2). For that, fix some values of x_3 and x_4. Then, we get a 2-D cross section of the 4-D space. A cross section of a 4-D convex polytope is a convex polytope. Changing the values of x_3 and x_4 will effectively change an additive term into the output neuron, and the corresponding cutting bend-hyperplane in the 2-D input subspace (x_1, x_2) will shift.

Fig. 4.7 The convex polytope carving mechanism generalizes to directed acyclic computation graphs and to RNNs

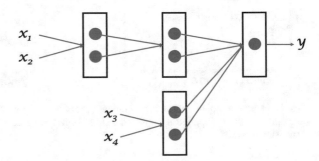

> *One of the reasons why AI works so well is because it subdivides the input space into exponentially many complex polytopes and fits it with a continuous piecewise linear function over these exponentially many pieces.*

4.6 Architectural Variations

The same geometrical carving is happening in any directed acyclic computation graph. In this book, we will encounter many architectural variations such as residual connections (e.g., in ResNet), skip connections (e.g., in U-Net, V-Net), fusion of computation paths, mixture of experts, multitask learning, and attention mechanism. The fusion can occur with addition operation or with concatenation operation.

The reader is encouraged to interpret these architectural variations in terms of carving. When it is a residual block

$$G(x) = F(x) + x,$$

we are adding finer details computed by F to the manifold. When it is a fusion with addition operation, $F_1(x) + F_2(x)$, we are combining the two carvings. For fusion with concatenation operation, we are moving to a product space. In particular, if it is a skip connection

$$G(x) = F_2(F_1(x)) \oplus F_1(x),$$

with concatenation, we informally say that there is flow of information from a lower layer F_1 to a higher layer G. We are working in the product space where we hope to achieve a better carving by G which takes into account both local, semantically weaker, and lower-level features computed by F_1 and global, semantically stronger, and higher-level features computed by F_2. For multitasking networks, the shared lower-level neurons first perform basic coarse carving, and then higher-level neurons for different tasks perform their own task-specific finer carving using hyperplanes which bend much more.

4.7 Attention and Carving

The attention mechanism modulates a feature by a weight which is dependent on the input. Therefore, we have a multiplication:

$$y = A(x)F(x)$$

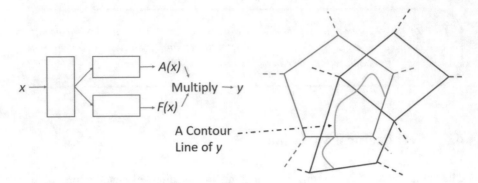

Fig. 4.8 The attention mechanism involves modulating a feature with another through the multiplication operation. $A(x)$ and $F(x)$ are piecewise linear functions on their respective set of convex polytopes, shown in red and blue, respectively. Their product is a piecewise quadratic function where pieces are convex polytopes which are obtained by the intersection of the red and blue polytopes. A contour line in such a piece will be a conic curve in two dimensions. In three dimensions, it will be a conic surface. This contour line will bend when the activation pattern changes and it enters an adjacent piece

where the attention weight $A(x)$ is dynamically computed based on the specific input and the feature $F(x)$ is accordingly modulated. If $A(x)$ and $F(x)$ are continuous piecewise linear functions, then their product is a continuous piecewise quadratic function. Suppose the set of convex polytopes for $A(x)$ is S_A and for $F(x)$ is S_B. The product $A(x)F(x)$ creates a piecewise quadratic function on a new set of convex polytopes S obtained by intersecting polytopes in S_A and S_B; see Fig. 4.8. Each polytope in S has an associated activation pattern of all neurons computing A or F considered together. Consider another example:

$$a = F_1(x)G_1(x) + F_2(x)G_2(x) + F_3(x)G_3(x),$$

$$y = [aF_1(x), aF_2(x), aF_3(x)].$$

If Fs and Gs are ReLU networks, then a is a piecewise quadratic function, and each component of y is piecewise cubic function. Such dot product operations occur in attention networks, capsule networks, and image similarity networks. For intuition about capsule networks, see Chap. 15.

For each polytope in the intersection, the function changes quadratically. This significantly increases the non-linearity of the overall function being carved out by the AI model. This nonlinearity is the attention mechanism, which is strengthening or attenuating of the original linear function (representing a feature) depending on the specific location of the input sample inside the polytope. We no longer have the property of straight lines being mapped to bending straight lines between the input space and the feature spaces. Now a line in one space maps to a non-linear curve in the other.

Suppose $y = f(x)$ is a quadratic function. The set of points in the input space for which $y = c$ for any constant c is a conic surface, which is its level set. In a 2-D input space, it is a contour line of y. On one side of the curve, $y > c$ and on the other $y < c$. When the activation pattern changes, we cross over to an adjacent polytope and the conic changes; see Fig. 4.8. It is a bending conic. We no longer have convex polytopes as pieces with flat faces on all sides. New added faces will be conic surfaces. Neurons in the layer preceding to the attention multiplication perform carving in the input space with bending hyperplanes. Neurons in the following layers do it with bending conics.

In self-attention used by transformers, the same process is happening to a higher degree. What if we were to repeat it again? Consider two such steps of attention:

$$y_1 = A_1(x) F_1(x),$$

$$y_2 = A_2(y_1) F_2(y_1),$$

where A_1, A_2, F_1, F_2 are the standard feed-forward networks and therefore they compute piecewise linear functions of their input. Since y_1 is a piecewise quadratic function, its composition with piecewise linear functions keeps $A_2(y_1)$ and $F_2(y_1)$ quadratic. The second multiplication is a product of two piecewise quadratic functions, making it a piecewise polynomial of degree four. The pieces correspond to different activation patterns. Note that pieces are mathematical objects with the *level sets of polynomials* as faces. The neurons subsequent to y_2 will perform carving of the input space with the bending level sets of polynomials of degree four. A convex polytope is a special case when the polynomial is of degree one and the level set is a flat face. We can generalize the carving theorem to include the multiplication neurons.

Carving Theorem 2 In a feed-forward ReLU network, in which some neurons multiply the outputs of other neurons, a neuron computes a continuous piecewise polynomial function where pieces have level sets of polynomials as their faces.

For a given hyperplane arrangement, if some of the hyperplanes are transformed to curves, that may potentially create even more complex arrangement of a larger number of regions, because the curve may bend around and intersect again. Just consider the case in which a conic curve intersects a straight line. There can be now two intersection points. Therefore, it only helps to have the multiplication neurons.

The proof of the theorem is the same as for the simpler linear version of the theorem. In a way, it is a proof by induction by building these functions layer by layer as we move from the input to the output. This provides the basis of the carving algorithm which we now write down for a general network:

1. Let x be the input to the network.
2. Perform topological sorting of the directed acyclic computation graph of the network. Let the sorted sequence be S.
3. Visit neurons in S in the sorted order. For each visited neuron y, do the following:

- For each activation pattern of the neurons preceding to y in S, on which y is dependent, perform the following:

  - Construct the polynomial function $P(x)$ being computed by y by adding, multiplying, and composing polynomials of the active preceding neurons in a forward pass.
  - Draw the level set $P(x) = 0$ in the region of the input space for which the activation pattern holds. On $P(x) > 0$ side of the level set, the neuron y is active. The other side is zeroed out by ReLU.

The above algorithm provides intuition into how AI creates its fantastic sculpture. We can consider other activation functions. For example, consider that ReLU has been replaced with the sigmoid function everywhere. ReLU discretizes the firing activity of neurons into regions. For the sigmoid function, all neurons are active, though some may be saturated toward 0 or 1. There is a single piece and a single highly non-linear function. Right at the output neuron, a level set will be created by a cutoff threshold to carve out the manifold for making the final binarized classification decision.

We end this section with a comment on a connection of AI with mathematics. Let us revisit level sets of polynomials. Consider a 2-D input space. Suppose a neuron is computing a polynomial of degree d. If $d = 1$, we have bending lines. If $d > 1$, we have bending conics. Now consider a 3-D input space, where we have surfaces. For $d = 1$, we have bending hyperplanes. For $d > 1$, we have bending conic surfaces. In an n-dimensional input space, we have bending high-dimensional conic surfaces. A level set is a solution to a single polynomial equation for a fixed activation pattern. The equation can be turned into an inequality to yield activation or inactivation regions.

How do these conic surfaces form the pieces of the piecewise polynomial function? Consider a neuron y. Fix the activation pattern A of the N preceding neurons on which y is dependent. A piece X is being subdivided by y. X is a solution to a system of N *multiple polynomial inequalities*, one for each neuron in A. The "sides" of X are the solutions when one of the inequalities becomes an equation. This happens when we cross over into an adjacent piece and a neuron in A just starts waking up or falling asleep. The "edges" and "corners" of X are the solutions when multiple neurons start flipping their state. The neuron y adds a new polynomial equation $y = 0$ to the system, dividing X into two subsets X_1 and X_2, for which $y > 0$ and $y < 0$, respectively. Note that when we move into an adjacent piece, the polynomial system has to be recomputed for a changed activation pattern A'.

Solving a system of a large number of polynomial equations with a large number of variables is one of the great challenges of the twenty-first-century mathematics. It is remarkable to see this connection between modern AI models and algebraic geometry. An open problem is to rework the bounds on the exponential expressive power for piecewise polynomial networks. And, what will happen if we add neurons for division of polynomials which leads to rational functions? In Chap. 15, we anticipate that Bayesian inference will play a role in the future of robust and interpretable AI. The probabilities computed by the AI models will get multiplied and divided by other probabilities. Such models will be understood at two levels. One will be at a high level in which we explicitly think in terms of a compositional process consisting of parts and whole. The other will be to understand how such methods will carve out highly complex manifolds.

> When we introduce multiplication in attention-based networks or capsule networks, then we carve out the manifold with more general pieces with curved faces. Each piece corresponds to an activation pattern. The AI model computes a polynomial for that activation pattern. Over the input space, it computes a piecewise polynomial function over curved conic-faced pieces. The study of the geometry of these modern AI models has links with great challenges in modern mathematics.

4.8 Optimization Landscape

4.8.1 Graph-Induced Polynomial

See Fig. 2.2. There are 15 computation paths connecting an input to the output. As an example

$$x_3 \to h_1 \to y$$

is one of these paths. Suppose the weight parameter of the edge from x_3 to h_1 is $w_{3,1}$ and of the edge from h_1 to y is u_1. Then, if h_1 is active, the input-dependent term

$$x_3 w_{3,1} u_1$$

will contribute a term in the computation of the output. Each path starting from an input will contribute such a term. If the bias parameter for h_1 is b_1, then the bias term $b_1 u_1$ will be computed which is input independent. Each path starting from a bias parameter will contribute such a term. There are four such paths. All these terms combine to form a polynomial.

We have seen that the output of a ReLU AI model is a continuous piecewise linear function of the input, where the linear coefficients are given by the activated part of the computation graph of the neural network based on the activation pattern invoked for the given N-dimensional input:

$$y = f(x) = \alpha_0 + \sum_{i=1}^{N} \alpha_i x_i.$$

The linear coefficient α_i for the input x_i will be the sum of the products of the weights over all the activated paths in the computation graph from the input x_i to the output y. If we were to keep x_i's constant, and vary the weights and biases, then

$$y = f(\theta)$$

is a polynomial in learnable parameters, where y is the output and θ is the set of learnable parameters. For classification, there is the additional step of squishing the output to the range $[0, 1]$. For regression, y is the answer.

The degree of a monomial term in the polynomial is the length of the path, and since weights are typically not shared between edges along a computation path, it is a multi-linear term of the form:

$$x_i w_{i_1} \ldots w_{i_d},$$

where the path of length d starts at the input x_i, and the maximum degree of a weight term is 1. Each monomial term has degree equal to or less than the depth of the AI model. For networks with the attention mechanism, we don't have multi-linear terms because the same weight may be shared by the feature network and the attention network, and later in the network, we will have a multiplication of the attention weight with the feature value.

In the previous section, the input was variable and the parameters were fixed. In this section, for a fixed input, we compute a high-degree polynomial of the learnable parameters. By composing the loss function and this polynomial, we get the loss surface over the parameter space.

> *For a fixed input, a ReLU AI model computes a graph-induced polynomial of the learnable parameters where the graph is dependent on the activation pattern. The composition of linear functions computed by the active ReLU neurons is the same as taking the product of the weights along the activated computation paths from the input or the bias parameters to the output, which contribute to the terms of the polynomial. The degree of the polynomial is equal to the depth of the model.*

4.8.2 Gradient of the Loss Function

The loss function depends on the output y and the ground truth G. For a minibatch, we take the aggregate of the loss functions of several inputs. Note that for the L^2 loss, the landscape will be a polynomial:

$$\mathcal{L}(\theta) = (f(\theta) - G)^2.$$

For any learnable weight $w \in \theta$,

$$\frac{\partial \mathcal{L}(\theta)}{\partial w} = 2(f(\theta) - G)\frac{\partial f(\theta)}{\partial w},$$

therefore, the gradient will be zero if either $f(\theta) = G$ or if the gradient of polynomial $f(\theta)$ is zero. For the cross-entropy loss

$$p = \text{sigmoid}(f(\theta)) = 1/(1 + \exp(-f(\theta))),$$

$$\mathcal{L}(\theta) = -G \log(p) - (1 - G) \log(1 - p).$$

Therefore, we have for any $w \in \theta$:

$$\frac{\partial \mathcal{L}(\theta)}{\partial w} = (p - G)\frac{\partial f(\theta)}{\partial w}.$$

See Exercise 6. Therefore, the gradient will be zero if either $p = G$ or if the gradient of the polynomial $f(\theta)$ is zero.

4.8.3   Visualization

See Figs. 2.3 and 4.9 for examples of loss functions over a 2-D parameter space. You can see numerous hills and valleys. In Fig. 2.3, the contour plot is also shown. Imagine yourself following the SGD algorithm and sliding down slopes till you find yourself at the bottom of a valley.

Fig. 4.9 Optimization landscape of a loss function over a 2-D parameter space. See also Fig. 2.3

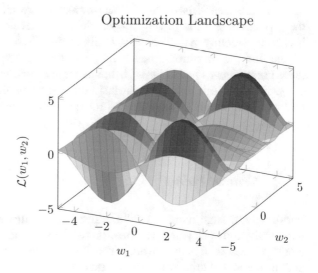

AI does the same with the big difference that it works in a high-dimensional space. Say there are 10 million learnable parameters in a CNN; therefore it is a 10 million-dimensional space, which is way beyond our visualization capability. In order to get a feel for the surreal landscape, one can choose a point in the trajectory of the SGD optimizer in this space and look at its vicinity. In particular, one looks at a planar cross section defined by two random, almost orthogonal, directions as the axes. The loss surface is plotted over this 2-D plane. See Fig. 4.10, which shows a result of this approach. Such appealing visualizations of loss surfaces in 2-D lead to interesting insights into their nature; see [129]. One can visually observe their rugged hills and valleys, and how different techniques such as dropout or batch normalization make the surface smoother or rougher. In fact, local minima at the bottom of the landscape seem to be connected by simple paths; see [55]. It is like taking a walk from one crater to another along a flat smooth path in a valley on an alien planet. This suggests exploring the surrounding valley by creatively varying the learning rate and taking an ensemble of the nearby models corresponding to low error values in a method known as fast geometric ensembling. By taking a running average of weights of these models rather than their predictions, in a method called stochastic weight averaging, one can approximate the ensemble by a single model. Applying this averaging during the entire SGD leads to less bumpy descent and smoother error curves.

Empirical experiments suggest that typical optimization landscapes have a simple geometrical structure. For example, if one were to plot the loss values along a line from the initial position θ_0 to the final position θ_f in the parameter space,

$$\mathcal{L}((1 - \alpha)\theta_0 + \alpha\theta_f),$$

by varying α from 0 to 1, then one gets simple, approximately convex curves; see [60]. Another way to investigate optimization landscapes is by restricting the number of dimensions in which the parameters are allowed to change. That is, the gradient descent is forced to occur in a d-dimensional subspace whose basis is chosen randomly in the complete D-dimensional space. Empirically one can vary d, and find the transition when SGD starts finding good solutions. Interestingly, these experiments suggest that d is not as large as one might expect. Furthermore, comparisons of this empirical intrinsic dimension of the landscape for different problems allow one to measure the relative difficulty levels of these problems; see [128].

4.8.4   Critical Points

Fortunately, a large number of dimensions means that there is almost always an opportunity to tweak some parameter(s) to continue the slide downward. There are many degrees of freedom available to improve the AI model. Recent empirical and theoretical work strongly suggests that local minima are relatively more concentrated toward the bottom of the optimization landscape whereas saddle points

Fig. 4.10 Optimization landscape of an actual loss function over a 2-D cross section of a high-dimensional space. This image of a loss landscape is created with data from the training process of a convolutional neural network. Specifications: Imagenette dataset, SGD-Adam, minibatch size of 16, batch normalization, learning rate scheduler, train mode, one million loss points, log scaled. This piece has been created through a multistage process that begins with the training of a deep learning network. Created by Javier Ideami. For more about the loss landscape project, visit https://losslandscape.com

are more numerous away from the bottom. Furthermore, most critical points are saddle points. Therefore during training we are likely to encounter local minima toward the end which correspond to good solutions to practical problems. Earlier in the training, it is highly likely that temporary slowing down of the training is because of a saddle point. A saddle point in Figs. 2.3, 4.9, and 4.10 will resemble a saddle on a horse. Gradients are zero, but it is neither a minimum nor a maximum. There are directions in which it is a minimum and in others it is a maximum. Its graph crosses the tangent plane at the point.

The presence of saddle points provides a justification for stochastic algorithms. The stochastic nature of SGD can assist in escaping saddle points by finding a direction which opens up on a new minibatch of data, that is, the gradient is nonzero in that direction and allows further descent toward the minimization of the loss function. The fluidity of the optimization landscape is helpful in escaping a saddle point. Consider the scenario in which ReLU neurons wake up or go to sleep on new minibatch samples, thereby having ripple effects on the bend-hyperplanes of neurons of higher layers they are connected to. We find ourselves in new convex polytopes and the loss surface changes. This noisy nature of SGD would help in escaping from a critical point.

And it is rare to encounter local minima. The way AI carves up the input space in terms of convex polytopes comes to our rescue. A new minibatch represents a new set of points. The carving bend-hyperplanes have to adjust to separate out the classes. It is hard to get stuck in a local minimum because there is room to tweak some bend-hyperplanes, which are as numerous as the number of neurons.

In high-school calculus, we learned that a local minimum of a function $f(x)$ satisfies

$$f'(x) = 0, \, f''(x) > 0.$$

The first derivative has to be zero and the second derivative has to be positive. For multivariate functions we replace the conditions with those on vectors and matrices:

1. The gradient vector has to be the zero vector for a critical point. The gradient vector is made of all first partial derivatives.
2. For the test of the second derivative being positive for a local minimum, we now require the Hessian matrix to be positive definite. The Hessian matrix is made of all second partial derivatives.

That is, the eigenvalues of the Hessian matrix have to be all positive at a local minimum. If the Hessian matrix is negative definite, then the point is a local maximum. If the eigenvalues are both negative and positive, then it is a saddle point. If any eigenvalue is zero, it wipes out the determinant of the Hessian matrix, making it a singular matrix, and then the critical point can be anything—local minimum, local maximum, saddle point, or none of these. In univariate case, this corresponds to the case when $f''(x) = 0$.

For random matrices, one can ask what the probability of all eigenvalues being negative is. For a special class of random matrices, known as Gaussian orthogonal ensembles, it can be shown that the probability of a critical point being a local minimum is very small. This result relates to a similar one for random polynomials, whose critical points are overwhelmingly saddle points. Thus it is highly unlikely that SGD will get stuck in a local minimum.

Moreover, saddle points are easy to escape early in the training when the loss is high. Toward the bottom of the landscape when the AI model is working well, they become harder to escape. Let's develop some intuition into this claim. Let index be the fraction of negative eigenvalues of the Hessian matrix. There is very little likelihood that all eigenvalues will be positive when the loss function is high because there are a large number of parameters to tweak. The index will be high. As the training proceeds, the index will reduce gradually, and when it reaches a local minimum, then it will become zero. When the loss is high, high-index saddle points dominate. As we descend the landscape, low-index saddle points become relatively more numerous. Intuitively, high loss implies that there are several directions in which one can descend down and there are many degrees of freedom to fix the errors and improve the model. As one approaches the bottom of the landscape, loss is low, the model is working great, errors are few, and there are few directions available to improve the model further. The optimization landscape has this interesting, very intuitive, layered structure. In addition, empirical results suggest that during and after training, most of eigenvalues of the Hessian matrix are zero making it singular. This implies there is an overall flatness in the layered optimization landscape.

In physics, spin glasses and Ising models have been studied to model systems in which particles interact with others under constraints. Their critical points have the same structure because there are many degrees of freedom to lower the energy of such physical systems. This provides a connection between AI and existing results. See [37, 42] to get an idea of how researchers are making progress in the analysis of optimization landscapes. We will continue to see more results in the future.

> *One of the reasons why large AI models can be trained is because of the special nature of high-dimensional optimization landscapes, in which there are very few local minima compared with the total number of critical points, making it unlikely that the training algorithm will get stuck in a local minimum.*

4.9   The Mathematics of Loss Landscapes

The previous section is a step in developing high level intuition about how an AI model gets trained in a high-dimensional loss landscape. This section is written for readers who are seeking advanced intuition and greater mathematical insight. The mathematics needed to understand such landscapes is challenging, yet it is possible to develop basic conceptual understanding.

4.9.1   Random Polynomial Perspective

We saw that the output of a ReLU AI model is a polynomial $f(\theta)$ and the critical points of the loss function $\mathcal{L}(\theta)$ coincide with those of the polynomial due to the chain rule of differentiation. Therefore we can look at the nature of critical points of polynomials and ask what is the probability of a critical point being a local minimum.

We will look at a special class of polynomials under certain specific assumptions to get an idea about the critical points of large polynomials. Consider a space of polynomials with degree two in two variables:

$$f(x, y) = a_1 x^2 + a_2 y^2 + a_3 xy + a_4 x + a_5 y + a_6,$$

where each a_i is an independent, centered, random normal variable. The variances are chosen to be multinomial coefficients, where the central terms, which can come from multiple sources, have higher variance. This space can be generalized to a space of polynomials with degree d in n variables:

$$f(x) = \sum_{|\alpha| \leq d} f_\alpha x_1^{\alpha_1} x_2^{\alpha_2} \ldots x_n^{\alpha_n},$$

where $\alpha = (\alpha_1, \ldots, \alpha_n) \in \mathbb{N}^n$ is a multi-index and $|\alpha| = \alpha_1 + \ldots + \alpha_n$. If we set the variances of the polynomial coefficients to the multinomial coefficients,

$$\frac{d!}{\alpha_1! \ldots \alpha_n! (d - |\alpha|)!},$$

then we get the space of random normal polynomials. Note that the variances indicate the number of sources which can contribute to the monomial term. For degree d, there are at most d sources, from each of which one can select a variable or choose not to select at all. Therefore, for the case $d = n = 2$, the term x^2 will have half the variance of the term xy, because for the latter there are twice as many ways to create the monomial. For the random normal polynomial space, using the big O notation from computer science, the probability that a critical point is a local minimum is

$$\Pr(\text{Local Minimum}) = O(\exp(-kn^2)),$$

where $k \approx 0.275$ (its exact value being $(\ln 3)/4$). This is quite amazing. The probability falls off exponentially as the number of learnable parameters n increases, which can be in tens of millions. The expected number of critical points increases exponentially at the same time:

$$O((d-1)^{(n+1)/2}),$$

with almost all being saddle points. Note that the degree d is the depth of the AI model. See [45] for details.

4.9.2 Random Matrix Perspective

The previous subsection presents the viewpoint from the random polynomial perspective. An alternative viewpoint is that of random matrices. In fact, the results about critical points of random polynomials are based on results from random matrix theory, in which real, symmetric, random, normal matrices called Gaussian orthogonal ensembles are used.

We mentioned the second-derivative matrix known as the Hessian matrix in the previous section. The eigenvalues of the Hessian matrix at any point in the optimization landscape capture the geometry of the loss surface at that point in terms of convexity and concavity of the local patch. The $n \times n$ Hessian matrix of a function $f(x_1, \ldots, x_n)$ is

$$H_{ij} = \frac{\partial^2 f}{\partial x_i \partial x_j}.$$

In AI it is a symmetric matrix because second partial derivatives are continuous. Consider a 2-D optimization landscape. Let the eigenvalues at a given point be λ_1 and λ_2. If $\lambda_1 > 0, \lambda_2 > 0$, then the Hessian matrix is positive definite, and the local patch around that point is concave up. If $\lambda_1 < 0, \lambda_2 < 0$, then the patch is convex down. If one is positive and the other is negative, it is concave up in one direction and convex down in another. If the point is a critical point, that is, the gradient is zero at that point, then the implication is that the critical point is a local minimum, a local maximum, or a saddle point, respectively, for the three cases. If any eigenvalue is zero, then we can't say much, and one really has to plot the function to draw conclusions. For an example, suppose $f(x, y) = 2x^3 + y^3 - xy$. Then,

$$\text{Hessian}(f(x, y)) = \begin{bmatrix} 12x & -1 \\ -1 & 6y \end{bmatrix}.$$

Note that at point (0,0), the eigenvalues are ± 1. Therefore, the patch at (0,0) is both convex down and concave up.

In order to see how eigenvalues describe the geometry around a critical point, consider the local Taylor series approximation around it:

$$f(x + \Delta x, y + \Delta y) = f(x, y) + \frac{1}{2}(\Delta x, \Delta y)^T H(\Delta x, \Delta y),$$

where H is the Hessian matrix. This is a second-order quadratic approximation of the landscape. Note that the first-order term drops out as the gradient is zero. Let $\lambda_1 > \lambda_2$ and let v_1 and v_2 be the corresponding orthonormal eigenvectors. If the vector $(\Delta x, \Delta y)$ is v_1, then the difference

$$f(x + \Delta x, y + \Delta y) - f(x, y) = \frac{1}{2}\lambda_1$$

is maximized, where we used the fact that

$$Hv_1 = \lambda_1 v_1$$

and that v_1 is of unit length. By moving along in this direction, if $\lambda_1 > 0$, then we have the steepest ascent. Similarly, the loss will decrease along an eigenvector with

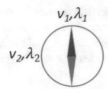

Fig. 4.11 If you are lost in the loss landscape with very poor visibility, the best thing is to have a compass which shows the eigenvectors and corresponding eigenvalues. That will give you some idea about the geometry of the local patch. If at some point, you notice that all eigenvalues in your eigencompass are registering positive values, then it will mean you have gotten trapped in a local minimum. How likely is this?

a negative eigenvalue. The sign of an eigenvalue determines whether moving along the corresponding eigenvector will be an ascent or a descent, and its magnitude determines by how much. See Fig. 4.11.

*** Exercise 12   Loss Landscape

Suppose a 2×2 Hessian matrix of an optimization landscape has eigenvalues λ_1, λ_2. Let v_1 and v_2 be the corresponding orthonormal eigenvectors. Suppose we move along a direction making an angle ϕ with v_1: how will the loss change?

We want to know what the probability is of all eigenvalues being positive. For Gaussian orthogonal ensembles, the probability is

$$\Pr(\text{All Eigenvalues are Positive}) = O(\exp(-kn^2))$$

where $k \approx 0.275$; see [44]. This result was used to obtain the result for random normal polynomials in the previous subsection. Gaussian orthogonal ensembles provide a model for the Hessian matrices; therefore it is highly unlikely that a critical point will be a local minimum. If we had modeled an eigenvalue being positive or negative as a Bernoulli process, like a coin toss, we would have concluded:

$$\Pr(\text{All Eigenvalues are Positive}) = O(2^{-n}).$$

This is exponential decay with n, but actual decay is significantly more rapid due to the squared term. Here is the intuition behind this amazing phenomenon. Coin tosses would have modeled the eigenvalues as independent random variables. In reality, they are dependent on each other in an intriguing way. A physical analogy is that of charged particles as in a Coulomb gas. The eigenvalues are restricted to a 1-D line and repel each other with a physical force. This repulsive force makes it highly unlikely that they will all get clustered on the positive side. In other words, *local minima are exceedingly rare*. SGD is likely to encounter only saddle points.

4.9.3 Spin Glass Perspective

A number of systems in physics, computer science, and mathematics are defined in terms of interacting components in which we have to minimize an objective function. These interactions make it difficult to satisfy all constraints which can conflict with each other. In physics, a spin glass or an Ising model consists of particles with couplings which indicate the mutual constraints. The goal is to find spins of the particles which satisfy these constraints as much as is possible. The couplings are fixed and the spins are variable. A configuration of spins has an energy that is known as the Hamiltonian. The more constraints are satisfied, the lower the energy. Since one cannot satisfy all of these constraints simultaneously, there is frustration. The goal is to reduce the frustration.

The energy landscapes of spin glasses and Ising models have been studied extensively due to their intriguing properties. Like in the optimization landscapes of AI models, they have many critical points, most of them being saddle points.

The AI analog of frustration is the loss function. Given a minibatch, SGD needs to update parameters so that the loss is reduced. The weights of the AI model are shared by many computation paths starting from different inputs, and this introduces mutual constraints. Therefore, an intuitive analogy exists between spin glasses and AI models.

An example of a spin glass is the Ising model:

$$H = -\sum_{i,j} J_{ij}\sigma_i\sigma_j,$$

where H is the Hamiltonian, the couplings J_{ij} are Gaussian random variables, and $\sigma_i = \pm 1$ are spins (degrees of freedom). If $J_{ij} > 0$, then the spins need to align. If $J_{ij} < 0$, then the spins need to be opposite. If $J_{ij} = 0$, then there is no interaction. The couplings are between neighboring particles and there is a 2-D spatial structure. This model can be generalized to a p-spin spherical model:

$$H = -\sum_{i_1 > \ldots > i_p = 1}^{N} J_{i_1 \ldots i_p}\sigma_{i_1} \ldots \sigma_{i_p}$$

where

$$\sum_{i=1}^{N} \sigma_i^2 = N$$

is the spherical constraint on real-valued spins. Now the Hamiltonian is the result of p particle interactions, all of which interact with all others, and there is no spatial structure in terms of neighbors. Note that H is a polynomial in spins, just as an AI model is a polynomial in its learnable parameters. The couplings for spin glasses take the place of the input variables in the AI model.

For spherical spin glasses, it is possible to perform a mathematical analysis and discover an interesting layered structure of eigenvalues. Recall that index is the number of negative eigenvalues of the Hessian matrix. The statistics of critical points of index k are related to the statistics of the $(k + 1)$-th smallest eigenvalue. What can be shown is as follows:

- At the bottom of the optimization landscape, there are only index 0 (local minima) critical points in a narrow band.
- In a band just above the bottom, there are local minima and saddle points with index 1.
- In a band above the second one, there are critical points with indices 0, 1, and 2.
- This layered structure continues with higher-index critical points appearing as the loss increases away from the bottom.

See [6] for details. Under certain assumptions, one can apply these results to AI; see [37]. The learnable parameters take the role of the p-spins, the loss is the Hamiltonian, and the couplings are given by the input minibatch. This can be seen by noticing the similarity between the monomial terms:

$$J_{i_1 \ldots i_p} \, \sigma_{i_1} \ldots \sigma_{i_p},$$

$$x_i \, w_{i_1} \ldots w_{i_d}.$$

The intuition is that when the loss is high, index is high, several eigenvalues are negative, and there are many directions in the saddle points to decrease the loss. Toward the bottom of the landscape, the AI model is working very nicely; thus there are few degrees of freedom to improve it further.

4.9.4 Computational Complexity Perspective

There is an interesting perspective from computer science on the nature of high-dimensional optimization problems. NP-complete problems are intractable problems to solve because their worst-case time complexity increases exponentially with the problem size.

An example is the 3-SAT Boolean satisfiability problem, which asks whether a Boolean expression in 3-CNF (conjunctive normal form) with M clauses and N Boolean variables can be satisfied. An example for $M = 3$ and $N = 5$ is

$$(x_1 \vee \overline{x_2} \vee x_3) \wedge (x_1 \vee x_3 \vee \overline{x_4}) \wedge (x_2 \vee x_4 \vee \overline{x_5}).$$

The decision problem is NP-complete. The optimization problem MAX-3SAT, which seeks to find the maximum number of clauses which can be satisfied, is NP-hard. Parameterize the 3-SAT problem with a parameter:

$$\alpha = \frac{M}{N}.$$

Clauses represent "constraints" and Boolean variables represent "degrees of freedom"; therefore the ratio α represents the difficulty of the problem. Using the analogy with spin glasses, clauses are like couplings and Boolean variables are like particles. Given an instance of 3-SAT, one can reduce it to an instance of the graph problem of maximal independent set. The vertices of the graph represent degrees of freedom, which will get colors, and edges represent constraints that no adjacent vertices get the same color. Therefore there is a direct analogy with spin glasses. Spin glasses are indeed closely connected to NP-complete problems; see [136].

In fact, the connection is much deeper, and this will further strengthen our belief that optimization in high-dimensional spaces is not intractable in practice. One can randomly generate instances of 3-SAT for different values of α and compute the fraction of expressions which are unsatisfiable. What can be shown is that as $\alpha \to 0$, most expressions are satisfiable. Intuitively, few constraints and many degrees of freedom make it easier to find a solution. As α increases, there is a phase transition when some expressions are satisfiable and some are not, and then for larger values of α, almost all expressions become unsatisfiable. This phase transition is similar to what has been observed in spin glasses; see [113].

Therefore, having many degrees of freedom leads to the solvability of hard problems. This is an important insight because in the case of AI, once we have fixed the size of the input, which represents "constraints," then by increasing the size of the network, which is the same as increasing the number of "degrees of freedom" of the AI model, one can satisfy these constraints. Note that it is well known that fast approximation algorithms can provide near-optimal solutions to very large problem instances of NP-complete problems, which provides empirical justification of the presence of easily found, near-optimal solutions in high-dimensional landscapes.

4.9.5 SGD and Critical Points

Since most of the critical points are saddle points in higher loss bands, we are unlikely to encounter a local minimum. How many saddle points will we encounter till we get stuck in a local minimum? We will work it out under the simplifying assumption that this can be formulated as a statistics question by modeling it as a geometric probability distribution. If the probability of a critical point being a local minimum is p, then the probability of encountering $t - 1$ saddle points followed by a local minimum during SGD is

$$\Pr(X = t) = (1 - p)^{t-1} p$$

with the expectation being

$$\mathbb{E}(X) = \frac{1}{p} = O(\exp(kn^2))$$

where $k \approx 0.275$. This is an astronomically large number when n is typically in the tens of millions. Therefore, it will be a long while before we encounter a local minimum. Note that p is not constant but changes as we descend down the landscape. Since p is smaller for earlier critical points, which belong to higher bands in the layered structure of the landscape, it is highly unlikely that we will get stuck in the higher bands.

Another question which arises is why SGD doesn't get stuck at a saddle point where the gradient is zero. The underlying reason is that SGD is a stochastic algorithm. It is unlikely that one will remain at a particular saddle point when new minibatches arrive. The landscape is fluid because the input changes and furthermore the underlying computation graph may change. Moreover, despite the gradient being zero, popular variations of SGD which keep a history of gradients can continue to move forward in the landscape. Saddle points may slow down SGD, but in practice they don't seem to stop it altogether.

*** Exercise 13   SGD

Read the paper [122]. Why does the noisy nature of SGD help in faster convergence to better solutions when compared with batch gradient descent in which the only minibatch is the whole training data? In classical ML, second-order methods such as the Newton method, BFGS, and the Levenberg-Marquardt algorithm have been used. Why don't we use them for training of AI models?

4.9.6 *Confluence of Perspectives*

In this section, we built advanced intuition into tools which are available to us to answer the question—why does an AI model get successfully trained? It is insightful to see a number of modern mathematical theories having rich interconnections:

1. Theory of random polynomials. Since a ReLU neuron computes a polynomial, one can apply results from this branch of mathematics, which depend on results from the theory of random matrices.
2. Theory of random matrices. Using Gaussian orthogonal ensembles, one can study statistics of eigenvalues of the Hessian matrix of the loss function, which also allows investigation of energy landscapes of spin glasses.

3. Spin glasses and Ising models. These models formulate mutual constraints which free variables have to satisfy. The Hamiltonian of these physical systems is a random polynomial of spin variables.
4. Computational complexity. Spin glasses are connected to NP-complete problems in computer science, for which it is intractable to reach the global minimum. Yet fast approximation algorithms provide near-optimal solutions to very large problem instances.

These different perspectives lead to the same insight that in high dimensions, the landscape has a rich mathematical structure which is amenable to algorithms such as SGD that have been fine-tuned to work very well in practice.

> *For advanced understanding of the optimization landscapes of AI models, one can employ several interconnected perspectives, such as the theories of random polynomials, random matrices, spin glasses, and computational complexity. Local geometry around a critical point is given by the eigenvalues, which repel each other making it highly unlikely that they will be all positive. There is a layered structure of loss landscapes in which high-index saddle points dominate when the loss is high, and it is easy to escape them with the SGD algorithm with very little likelihood of getting stuck in local minima. Large number of degrees of freedom in these systems implies that constraints can be easily satisfied to obtain a good solution.*

4.10   Distributed Representation and Intrinsic Dimension

In the book, we have informally defined a distributed representation as a firing pattern of neurons; see Fig. 4.12. See [87] for a discussion on the benefits of distributed representations. In Part II of the book, which deals with applications, the reader will notice that a number of AI solutions across different application areas are based on the *encoder-decoder architecture*, in which the encoder builds a representation R of the input, which is then used by the decoder to produce the output. That is, we typically have the following:

$$R = E(x),$$
$$y = D(R),$$

where E is the encoder, D is the decoder, and (x, y) is the input-output pair. One of the primary goals of AI is to learn good representations. This is called *representation learning*. In general, if the AI model has an underlying computation graph, then one can define a graph cut as a partitioning of the graph into two parts, one of which has the input and the other has the output. One can refer to the graph on

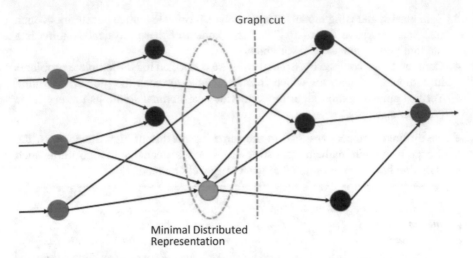

Fig. 4.12 A deep neural network is shown as a directed acyclic computation graph. It builds a distributed representation of an input, in a latent semantic space, as the firing pattern of the neurons for that input. Since there is causal dependence between the firings of neurons, the firing pattern of a subset of neurons can serve as a representation. The activation values of the green neurons in the figure yield a minimal-size representation of the input. The graph cut conceptualizes the AI model as an encoder-decoder architecture. A useful strategy in AI has been to build robust representations of the input for certain tasks and then attach customized task-specific decoders to these pre-trained encoders to fine-tune the model further

the input side as the encoder, and to the graph on the output side as the decoder. The vertices connected to the cut-set, the set of crossing edges, on the input side are a representation of the input. The representation with the minimum number of vertices is the most compact representation, and its size is an estimate of the intrinsic dimension of the target manifold or the function which the model is trying to approximate. Ingenuity in the design of many AI solutions lies in the proper choices of the encoder and the decoder.

This is related to the *manifold learning hypothesis*. Consider a 1-D curve in the 2-D space. As long as it is not a space-filling curve, its intrinsic dimension is 1 though it is embedded inside a 2-D space. The intrinsic dimension of a fractal curve will be its Hausdorff dimension. The manifold learning hypothesis states that the manifolds corresponding to typical classes in AI have a dimension d which is much smaller than the input dimensionality N, that is, $d \ll N$. The intrinsic dimension of the manifold, which can twist and turn around in the surrounding N-dimensional space, corresponds to the number of characteristic latent features of the class. An AI model discovers these features, which constitute the latent representation.

Being able to converge to a low loss in the optimization landscape without encountering local minima means that one is building good representations of the input space. Why does the built representation work well for the test data? AI models have high expressive power, and given enough data and ingenious design, they can avoid overfitting as discussed in Chap. 5, which leads to their ability to generalize to unseen data. This is another insight which is needed to completely understand the success of AI. Creatively designed models with high expressive power (carving capacity) when given an adequate amount of data work very well.

4.11 Chapter Summary

There are two reasons why AI works well. The first reason is that it has an exponential expressive power in subdividing the input space into convex polytopes, each of which is fitted with a linear function. The whole fitted function is a continuous piecewise linear function with exponentially many pieces. This allows AI models to approximate complex functions and complex manifolds. Each neuron contributes its bend-hyperplane cut to the input space, which further refines the polytopes carved by the previous layers. This analysis can be generalized to CNNs and RNNs. In the case of a CNN, the input space is the sliding window space, and the carving has translation invariance and robustness to local perturbations due to pooling layers. In the case of an RNN, the computation graph is like a directed acyclic graph with weight sharing, and the same carving process works. When we add neurons which perform multiplication, the AI models fit piecewise polynomial functions where pieces are formed by high-dimensional conic sections.

The second reason why AI works well is because the optimization landscape is a high-dimensional one in which almost all critical points are saddle points rather than local minima. The SGD algorithm, which makes the optimization landscape fluid, assists in escape from the saddle points. Empirical and theoretical works on high-dimensional physical systems such as spin glasses and Ising models have pointed to the concentration of local minima toward the bottom of the landscape. Several perspectives from mathematics can be applied to get insights into the nature of optimization landscapes of AI models. These include the theories of random polynomials, random matrices, spin glasses, and computational complexity. By converging to a low-loss solution, SGD is able to build a good representation of the input.

Problem Set 4

1. Why does a hyperplane bend when it moves from one activation pattern region to another?
2. Given an example of a 2-D polytope. When will a polytope be convex?
3. Show that $n^3 + n^2 = O(n^3)$. Is it also $\Theta(n^3)$? What about $n^3 - n^2$?
4. Suppose the input to a neural network is d-dimensional and there is a single output neuron. Assume that the expressive power for a network with L hidden layers, each having n neurons, is exactly cn^{dL} for some constant c. Consider a shallow network with one hidden layer with n_1 neurons. Consider a deep network with L hidden layers each with n_2 neurons. What should be the value of n_1 so that their expressive powers are equal? Compare the number

of parameters when their expressive powers are equal. Choose $d = 100$, $n_2 = 100$, $L = 50$ to calculate the exact numbers.

5. See Fig. 2.2. In total there are 15 computation paths starting from the input. Suppose we add a higher-level hidden layer with five neurons. What is the new number of computation paths? Suppose h_2 becomes inactive. Then, how many paths will be left?

6. Consider the following 2-D feed-forward fully connected ReLU network with two hidden layers:

$$r_1 = \text{ReLU}(w_1 x_1 + w_2 x_2 + b_1),$$

$$r_2 = \text{ReLU}(w_3 x_1 + w_4 x_2 + b_2),$$

$$s_1 = \text{ReLU}(w_5 r_1 + w_6 r_2 + b_3),$$

$$s_2 = \text{ReLU}(w_7 r_1 + w_8 r_2 + b_4),$$

$$y = w_9 s_1 + w_{10} s_2 + b_5$$

Suppose the neurons r_2 and s_1 are active and neurons r_1 and s_2 are inactive. Write the polynomial being computed by the network. Assume that the inputs x_1 and x_2 are fixed and the weights and biases are variable. What is the degree of the polynomial? Now assume that the inputs are variable and the weights and biases are fixed. What is the degree of the polynomial?

7. Why is the Hessian matrix important in machine learning?

8. For each of the following functions, indicate if there is a critical point at the specified value. If it is a critical point, indicate the type of critical point:

   (a) $\sin x$ at $x = 3\pi/2$.
   (b) $x^3, x^4, -x^4, x^5$ at $x = 0$.
   (c) $x^2 + y^3$, $x^2 - y^2$ and $x^2 + y^2$ at $(x, y) = (0, 0)$.

9. Define eigenvalues and eigenvectors. What are the eigenvalues and eigenvectors of a 2-D non-uniform scaling matrix with different scaling factors in x and y directions?

$$\begin{pmatrix} s_x & 0 \\ 0 & s_y \end{pmatrix}$$

10. Consider three particles A, B, and C in a 2-spin glass. Suppose the constraints are $J_{A,B} > 0$, $J_{B,C} < 0$, and $J_{A,C} > 0$. Each particle can take either $+1$ or -1 spin value. Find spins which will satisfy the constraints.

11. Suppose two polynomials $P(x)$ and $Q(x)$ have degrees d_1 and d_2. What are the degrees of the product $P(x)Q(x)$ and the composition $P(Q(x))$? Suppose inspired by self-attention in transformers, which uses query, key, and value functions, you design a novel way to perform NLU tasks as follows:

$$q_1 = Q(x),$$
$$k_1 = K(x),$$
$$v_1 = V(x),$$
$$q_2 = q_1 A_1(x) A_2(x),$$
$$k_2 = F(k_1) A_3(q_2),$$
$$v_2 = q_2 k_2 v_1.$$

Suppose all seven functions shown in the upper case are the standard ReLU feed-forward networks. What are the degrees of the six piecewise polynomial functions q_1, k_1, v_1, q_2, k_2, and v_2?

12. Let $P(x) = ax^2 + by^2 - 1$. Indicate the geometry of its level set $P(x, y) = 0$ for the following cases:

(a) $a = b, a > 0, b > 0$.

(b) $a \neq b, a > 0, b > 0$.

(c) a and b have different signs.

Chapter 5
Novice to Maestro

> When you're experimenting you have
> to try so many things before you
> choose what you want, and you may
> go days getting nothing but exhaustion.
>
> Fred Astaire

> All life is an experiment. The more
> experiments you make the better.
>
> Ralph Waldo Emerson

The same way a sculptor creates a piece of art, an AI/ML model approximates a manifold. AI/ML models differ in their abilities. Some perform very simple approximation, and some can handle highly complex manifolds. Those which are simpler are said to have *high bias*. Those which are capable of complex approximations have less bias and many degrees. They can carve in flexible ways depending on the training data. They are said to have *high variance*. There is a trade-off here because a high-bias method will typically have lower variance and vice versa. A less skilled sculptor carves a statue crudely with limited tools, whereas a more skilled sculptor who is equipped with finer tools does a sophisticated job in capturing the subtle features and curves. A skilled sculptor can come up with many "variations" to approximate what he/she has been told about the target object.

Despite the analogy with the art of sculpture, the way AI carves out manifolds is different from the way Rodin and Michelangelo worked. The latter were geniuses who knew exactly how to produce objects of beauty with perfection. AI has to learn a lot and go through a careful process of training.

© The Author(s), under exclusive license to Springer Nature Switzerland AG 2021 101
S. Dube, *An Intuitive Exploration of Artificial Intelligence*,
https://doi.org/10.1007/978-3-030-68624-6_5

5.1   How AI Learns to Sculpt

An AI is a novice who does not know how to chisel properly without being taught. It learns over time by looking at examples. It does so by improving on its previous effort, making fewer and fewer mistakes and gradually getting it right. It has to learn how to use its budget of linear regions judiciously so that as few mistakes are made as possible.

This is the standard training process of supervised machine learning for classification and regression. Training happens in epochs, as we saw in Chap. 2. To recapitulate, in an epoch, one pass is made over the training set, which is divided into minibatches of training samples. The AI figures out how to slightly modify its chiseling so that fewer mistakes will be made, as indicated by the specified loss function. A gradient step is taken in the direction of minimization of the loss function through the process of Stochastic Gradient Descent (SGD). The process is repeated minibatch after minibatch, epoch after epoch, till the desired accuracy is achieved. The gradient step which is taken becomes smaller with time, because the gradients get smaller, and moreover it is determined by the learning rate, which becomes smaller with time. In the initial epochs, AI is a novice and has to learn fast by tweaking on its previous effort more aggressively. Toward the end, it is almost a maestro and its sculpting skills have to be improved only slightly.

---

* Exercise 14   Training

Suppose you are training an AI model using SGD and you notice that its accuracy has plateaued. What hyperparameter can you tweak as the first step to potentially fix the problem?

5.1.1   Training Data

We evaluate the works of Michelangelo by their sheer beauty. The sculpture carved out by an AI is evaluated based on how well it performs in practice. On the training set, it is quantified by the training error. Since these training samples are visible to the AI, with judicious choice of the neural network architecture and of the training protocol, it can do a pretty good job in reducing the training error. However it can't really see the whole manifold and the real evaluation will be on unseen data. This is test error or generalization error. Learning is called robust if it generalizes well to the rest of the points in the sculpture. If we have low training error, but high test error, then it implies that we have overfitted to the training set. If the training error itself is high, we are underfitting. More discussion about these concepts will follow in this chapter.

The training process is more than running SGD. To set up SGD, we need hyperparameters which determine the training protocol. The minibatch size, number of epochs, learning rate, momentum, etc. have to be properly initialized. The loss function can be modified and a regularization term added. We will discuss regularization in Sect. 5.1.4. The size of input, the size of training data, and data augmentation have to be decided. One can choose to perform end-to-end learning or break up the system in components. Bias and variance have to be evaluated, and overfitting and underfitting problems have to be diagnosed. The architecture of the AI model can be modified and training data increased. Increasing the model size will increase the running time of training as well as inference. Therefore time profiling of different models is needed. The input can be preprocessed and enriched. And constantly we have to analyze errors and take steps to reduce them.

How are we going to experiment when there are so many choices? We can't use the test set for tuning of hyperparameters because we will overfit to it. Therefore in practice we put aside a separate set called the development or validation test in order to evaluate our choices. You are allowed to use the development set as an AI debugging set, the results on which you can manually and visually examine in depth. If the development set is big, you should use only a small portion of it for in-depth analysis and use the remaining portion to get high-level accuracy numbers, in order to avoid overfitting to it. Here are the three fundamental datasets:

Training Set   Set of data samples used by SGD.

Development or Validation Set   Set of data samples used to guide and evaluate experiments on the training set.

Test Set   Set used to evaluate the final model. This evaluation should not feed back to decisions which fine-tune the training process. If in-depth analysis of errors on the test set is done in order to guide the training, then the test set should be included in the development set, and a new test set should be obtained.

Often the training set is expanded by transforming each training sample. A cat's image can be cropped, resized, flipped, rotated, and dithered (dithering is addition of noise). This is called data augmentation, and it can improve the performance of the AI models.

5.1.2   Evaluation Metrics

For evaluation, the chosen metric should reflect your use case. For classification, Receiver Operating Characteristic (ROC) curves are popular. For information retrieval and object detection, where one has a very large number of negative examples, one uses Precision Recall (PR) curves. One can take the area under these curves as a single evaluation metric. These are called Area Under Curve (AUC) and Mean Average Precision (mAP), respectively, for the ROC and PR curves. An operating point on the curve can be chosen determined by requirements of the AI-based product. Precision and recall numbers can be combined by taking their harmonic mean, known as the F1 score. Often the single-value metric of accuracy suffices to judge the performance. For multi-class problems, the top-k error is the fraction of time when the correct answer does not show up in the top k results sorted by probabilities. Another natural way is to reduce a multi-class problem to a collection of binary classification problems by considering one-versus-rest metrics and plotting curves for each case.

All evaluation metrics depend on counting of errors and taking ratios; therefore numbers are in the range [0, 1] and can be expressed as percentages. Each answer of the AI model is labeled as true positive, false positive, true negative, or false negative, depending on what the ground truth is and what the predicted label is.

	Positive prediction	**Negative prediction**
Positive label	True positive	False negative
Negative label	False positive	True negative

An AI model for classification depends on a cutoff threshold in order to output a classification decision. If the computed class probability is over the threshold, then it is a predicted positive class member. The threshold is incrementally increased from low to high to produce ROC or PR curves. The True Positive Rate (TPR) is the ratio of the number of true positives to that of labeled positives. TPR is known as recall or sensitivity. The False Positive Rate (FPR) is the ratio of the number of false positives to that of labeled negatives. It is the same as 1 minus specificity. The precision (positive predictive value) is the ratio of the number of true positives to that of predicted positives. 1 minus TPR is the Type-I error in statistics, and FPR is the Type-II error. For regression, Mean Squared Error (MSE) or mean absolute deviation (MAD) is used. The goodness-of-fit metric R^2, known as the coefficient of determination, is frequently used for linear regression.

The ROC curve plots FPR (x-axis) vs TPR (y-axis), and the PR curve plots precision (y-axis) vs recall (x-axis), parameterized by the cutoff threshold $t \in [0, 1]$:

$$\text{ROC}(t) = (\text{FPR}(t), \text{TPR}(t)),$$

$$\text{PR}(t) = (\text{Recall}(t), \text{Precision}(t)),$$

that is, we judge x to be in the positive class when

$$\Pr(x \in \text{Positive Class}) \geq t.$$

The ROC curve goes monotonically upward, but the PR curve can fluctuate while having an overall downward trend. See [43] for the relationship between these widely used evaluation curves. A curve dominates in ROC space if and only if it dominates in PR space. PR curves are used in the settings when it is difficult to compute FPR because of an extremely large set of ground-truth negatives. An example is object detection in computer vision, where the number of image regions not containing the target object is very large. Another example is web search, in which for a given query there are a few relevant results and an astronomically large number of irrelevant results.

Metrics are often customized to particular domains. Examples are perplexity for language modeling in NLU, bilingual evaluation understudy (BLEU) for machine translation in NLU, and Normalized Discounted Cumulative Gain (NDCG) in web search. Lastly, the running time at the inference should be considered as part of the evaluation process.

The training process involves manually looking at the errors which the AI model is making. This enables one to debug the whole process and identify the problems, which in turn leads to their solution.

5.1.3 Hyperparameter Search

Finally, some of the hyperparameters used in training AI models can be fine-tuned systematically using approaches which vary from simple ones to complex ones. In the randomized-grid approach, one uses uniformly placed values on the hyperparameter grid, with random noise added to avoid repeating the same value. Thus you may try all learning rate values [0.1, 0.2, 0.3, 0.4]. If you have momentum as another hyperparameter, you can try [0.8, 0.9]. For both hyperparameters, you can try the randomized grid:

$$(0.10, 0.81), (0.20, 0.79), (0.30, 0.82), (0.40, 0.77)$$

$$(0.11, 0.89), (0.19, 0.91), (0.32, 0.93), (0.38, 0.88).$$

In Bayesian hyperparameter search, one applies Gaussian process regression (GPR) or Kriging, a form of Bayesian inference, to predict how well the training will work on new hyperparameter combinations, with a trade-off between exploration and exploitation. We will discuss GPR later, in Chap. 8. One can also employ evolutionary algorithms with effective searching and pruning techniques.

5.1.4 Regularization

Suppose you are aiming for something in life. You can take SGD-like steps toward your life goal and accomplish what you have been wanting to do. In the real world, it takes money and time to reach some goals, and there are constraints on these because they are not unlimited. You have to work toward the optimum under these constraints. Optimizing under constraints is called regularization. In the introductory calculus, this is known as constrained optimization. The goal is to optimize $f(x)$ with a constraint $g(x) = C$. The search space gets limited by

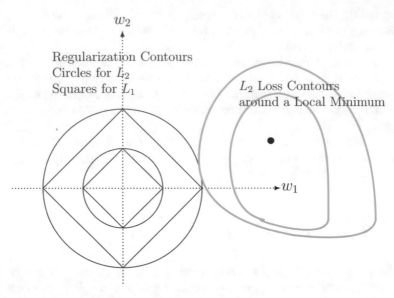

Fig. 5.1 We want to optimize a non-convex function $f(x_1, x_2)$ defined over a 2-D parameter space. Its contours are shown in green and enclose a local minimum. Adding a regularization term $g(x)$ means we are minimizing $f(x) + \lambda g(x)$, where λ is the Lagrange multiplier. If $g(x)$ is the L^2 norm of the parameters, then its contour plots are blue circles, for varying values of λ. A local minimum will be where a green contour touches the blue circle for the given λ. For the L^1 norm regularization, a green contour has to touch a red square

regularization. Constrained optimization can be solved by the Lagrange multiplier method, which optimizes a new function:

$$f(x) + \lambda g(x).$$

In AI, we do exactly the same to reduce overfitting.

See Fig. 5.1. These are constraints on the magnitude of parameters, and the goal is to keep them small. Consider this as our prior belief that simpler models are better. The L^p norm of the parameter vector is used as the constraining function $g(x)$. Now the optimal value is found where the contour of $f(x)$ touches the constraining contour of $g(x)$, which will be a circle for the L^2 norm and a rotated square for the L^1 norm. The latter is used to promote sparsity because the contours of $f(x)$ are more likely to touch the corners of the squares.

Given lots of data, we won't need any regularization. In practice, regularization does help because it avoids overfitting and reduces variance.

Let us build further insight about the regularization. Suppose there is a convolutional layer followed by a batch normalization layer. We ask the question why the batch normalization layer, which normalizes the scale of the output of the convolutional layer, and a regularized loss do not shrink the weights of the convolutional layer to zero during the execution of SGD. SGD computes a gradient vector ∇w, scales it with the learning rate, and adds its negative to the current parameter vector w in the parameter space. In a high-dimensional space, the two vectors are likely to be orthogonal, and an infinitesimal gradient step is effectively an infinitesimal rotation of w. The batch normalization and a regularized loss will make the rotation steps spiral inward, whereas a counteracting learning rate will make them spiral outward, thereby providing a balance between the two forces.

5.1.5 Bias and Variance

Every ML model comes with its bias and its variance. Bias quantifies what error we can expect because of the inherent limitations of the chosen ML model. A linear model gives us only one hyperplane, and it will be hard to approximate a complex manifold with it. Thus it has high bias. A DNN with its exponentially many convex polytopes will have very low bias. At the same time, if we were to choose a different set of training examples, the linear model will be more stable than the DNN. Statistical fluctuations in the training set get amplified more in the case of a DNN. Thus it has more variance. Therefore if we were unlucky in having training data which is not representative of the actual data, it will be reflected in a higher error for a DNN.

The mathematical definition of bias and variance is in terms of expectations. Fix the training data size to N. Train the AI model with many training datasets. Evaluate the test error for each choice. The average of the test errors gives the bias, and the variance of the errors gives the variance. The bias and variance tell us how statistical fluctuations of the training sets determine the test error.

Unfortunately, we have only one training set and one test set. Therefore there is only one test error. As a practical rule of thumb, we can estimate the bias with its lower bound, which is the training error. If the lower bound goes up, the bias will go up. The generalization error, which is the difference between the training error and the test error, is due to the random chance of picking up the particular training set. We can estimate variance with this difference.

Looking at these two estimates of bias and variance, you can quickly characterize your AI model. If it has high bias or high variance, it is time to mitigate the problem. High bias and low variance indicate underfitting, and low bias and high variance indicate overfitting. Low bias and low variance are the ideal situation. High bias and high variance mean something is just not right.

	Low bias	High bias
Low variance	Ideal	Underfitting
High variance	Overfitting	Undesirable

High bias and underfitting imply that the AI model is "underqualified" to do the job; an analogy is that you want your small child to help you with your AI conference presentation. To fix high bias, just make the AI model more qualified; in other words, increase its capacity by adding more neurons and layers. In our quite informal analogy, raise your child into a fine productive adult striving for excellence in challenging tasks. Another possibility is that you can decrease regularization at the risk of increasing variance. Regularization is a constraint on expression of full capacity, so decreasing the constraint will increase its capacity. Regularization is like hiring a highly qualified person but asking him/her to perform jobs below his/her qualifications.

High variance and overfitting mean that AI model is "overqualified" to do the job; another quite informal analogy is that you want a pet elephant to carry your AI book for you. To fix high variance, make the task more challenging. Give the model more data so it works hard to use its excess unused capacity to learn from it. Increase regularization to increase the constraint on capacity; however be mindful of the fact that since it reduces the capacity, therefore it may lead to high bias. The model size can be reduced, but first see if other methods work. Overfitting can be also reduced by early stopping. Stop the training as soon as the error on the development set starts increasing. In AI, a technique called dropout is used to reduce overfitting, which randomly drops neurons during training to reduce correlations between them and to nudge them toward learning robust specialized features.

To reduce both bias and variance, you have to go back to the whiteboard. Brainstorm, look at the problem from a fresh perspective, redesign your system, add new features, modify existing ones, and come up with innovative architectures. This works best when there are no obvious solutions such as increasing training data or model size. Keep in mind that you can probably never reduce the test error to zero. The best possible error rate is called the Bayes error rate. This is irreducible error which we can never get rid of, even with the best classifier. This is because there is always some source of uncertainty. If we know perfectly and there is no random noise, then the Bayes error will reduce to zero. This inevitable error can sometimes be measured by asking human experts to make calls. If experts can't achieve zero error, an AI won't be able to. Finally, observe that the total error is the sum of the squared bias, variance, and irreducible error.

---

* Exercise 15   Bias-Variance

Write mathematical definitions of bias and variance.

5.1.6   A Fairy Tale in the Land of ML

The bias and variance trade-off has a fairy-tale analogy; see Fig. 5.2. In an enchanting land faraway, there lives a beautiful princess courted by many princes from different places. Different princes have different abilities as given by their

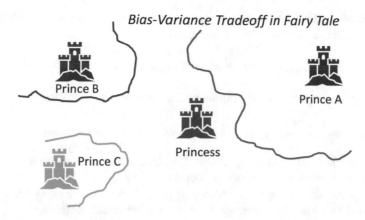

Fig. 5.2 Bias-variance trade-off in a fairy-tale setting. The princess represents the ground truth. Every (x, y, z) point in the 3-D landscape is a 3-D vector of predicted values for a test set with three samples. Each prince is an AI model which wants to reduce the total error. Fix the training set size to some N. Green Prince C has high bias and low variance as indicated by the green line, where he can land on after going through a successful training on a randomly chosen dataset. The final distance to the princess will be the total error. Dark Red Prince B has medium bias and medium variance. Deep Blue Prince A is most resourceful and can approach closest to the princess due to his low bias. At the same time, he suffers from high variance

biases and variances, and their goal is to get as close to the princess as possible to declare their intention for marriage.

High-bias and low-variance princes have limited resources. Their philosophy is rather simple, and they are able to cover a limited area in the land as they approach the palace where the princess lives. At some point they get stuck, as the terrain is rough and challenging. Their variance is low, which means they pretty much reach the same final spot even if they encounter different sets of circumstances on their journey, which correspond to the training process. A final location is the final evaluation of a particular training session on a test set. In contrast, low-bias and high-variance princes are more resourceful and are capable of overcoming ravines and waterfalls and other dangers on the way. They are capable of reaching closer to the princess, but their final location is less predictable. They sometimes land here and sometimes there depending on luck and circumstances. Low-bias and low-variance princes have the greatest chance of winning the princess's heart as they get closest because they are more immune to precarious circumstances. They reach the princess to profess their matrimonial intentions, but none can overcome the final irreducible distance due to unexplained uncertainties. We can speculate that only if it is a story of true love as we often witness in Hollywood fairy tales, which comes with absolute certainty, will the Bayes error be zero.

What will be the closest mathematical justification of the above fairy-tale analogy? In the fairy tale, the landscape is in a 3-D world. Each location (x, y, z) is a learned function's evaluation on three unseen examples, though in the actual definition it should be over the entire distribution in an infinite-dimensional space or at least on a very large test set. A prince's journey can be described as a sequence of positions obtained over epochs of the training session which indicate the intermediate evaluations on the 3-D test set:

$$P_1 \rightarrow P_2 \rightarrow \ldots \rightarrow P_{\text{final}},$$

as the training converges to an optimal point. The princess's location

$$G = (g_x, g_y, g_z)$$

is the true function value, the absolute golden truth. The number of training examples is fixed to N. Each prince, corresponding to an ML model, can land at different places depending on the choice of the training set. The expected final spot is obtained by taking the expectation, in the mathematical sense, over all possible training sets of size N. The distance between a prince and the princess is the total test error, which will be the sum of the squared bias, variance, and irreducible error. The irreducible error reflects unknown and unobserved processes which affect the input-output mapping.

Are we saying that it is impossible to achieve low bias and low variance? Not really: as we discussed earlier, with ingenious design, and as we will shortly see with enough data, we can achieve that. Therefore, a prince with adequate training and with ingenuity on his side can indeed meet the challenge in our imaginary fairy-tale setting. In fact, this is one of the reasons why AI has been successful.

> *The process of training an AI model starts with organization of data into three sets of training, development, and test data. Techniques to reduce overfitting include regularization, dropout, and evaluation of the development set. Bias-variance analysis is required to diagnose underfitting and overfitting problems in the training.*

5.2   Learning Curves

Learning curves give us a holistic view of errors as we change the training data size n. A learning curve is a plot of the training error and the test error as they vary with n. They allow us to predict whether it is worth getting more training data or not. And they provide an estimate of bias and variance for a particular value of n and how they will change as we increase n.

Let $\mathcal{E}_{\text{Train}}$ be the training error and $\mathcal{E}_{\text{Test}}$ be the test error. Let s be the SGD iteration number, H be the expressive power (capacity) of the AI model, and n be the number of training data samples. Then we have the dependence:

$$\text{Error During Training} = \mathcal{E}(s, H, n)$$

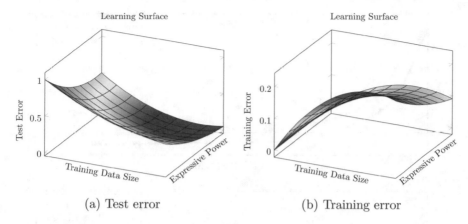

(a) Test error   (b) Training error

Fig. 5.3 Learning surfaces. See Fig. 5.4 for their cross sections, which are known as learning curves. We show the test error and the training error as a function of N, the number of training data samples, and of H, the expressive power (model complexity) of the AI model. With increasing N, the test error decreases and the training error increases. As we increase H, the training error decreases, and the test error initially decreases and then starts increasing when the model starts overfitting. (**a**) Test error. (**b**) Training error

for both the test error and the training error.

See Fig. 5.4. At the top, we fix H and n, and we show how the two errors depend on s. Consider the lowest test error $\mathcal{E}_{\text{Test}}(s)$ before it starts increasing in the overfitting region. Notice that $\mathcal{E}_{\text{Train}}(s)$ keeps on decreasing.

Call these errors the final errors on the trained model:

$$\text{Error at the end of Training} = \mathcal{E}(s^*, H, n),$$

where s^* is the optimal number of SGD steps without overfitting. We will drop s^* now and just work with

$$\text{Final Error} = \mathcal{E}(H, n).$$

This function is called the learning surface, as shown in Fig. 5.3. It is hard to see what's really going on there; therefore normally $\mathcal{E}(H)$ is plotted with fixed n, and $\mathcal{E}(n)$ is plotted with fixed H. We fix n and vary H in the middle of Fig. 5.4. At the bottom of Fig. 5.4, we vary n and show curves for two different values of H. These are called learning curves.

In the middle of Fig. 5.4, when H is varied, initially there is underfitting because the AI model has high bias. Then, there is a region where the model has optimal expressive power. If the capacity of the model is increased further by adding more layers, then overfitting occurs and the model suffers from high variance.

At the bottom of Fig. 5.4, we show markedly different behavior of $\mathcal{E}_{\text{Train}}$ from $\mathcal{E}_{\text{Test}}$. The training error increases, whereas the test error decreases with increasing

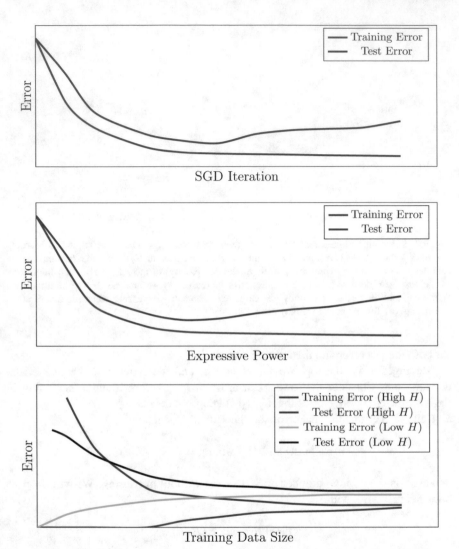

Fig. 5.4 Learning curves. The bottom two plots are cross sections of the learning surface. In the top two graphs, notice that overfitting occurs when the test error eventually starts increasing after initially decreasing. In the bottom figure, the model with high expressive power (high H) overfits when there is inadequate data. Given enough data, it gives a smaller test error, whereas the model with low expressive power underfits. Ingenious design of an AI model ensures that its high carving capacity (expressive power) is most amenable to the problem. That's why AI has been successful

n. When the expressive power is high and the training data is limited, we have high variance, and we overfit the high-H model to the training data, leading to high test error. The overfitting problem can be fixed by increasing n, as you can see the red curve drops below the black curve after a particular value of n. For large n, the low-H model is underfitting and has high bias. By switching to high H, we fix the high-bias problem and lower the test error to the red curve.

By plotting learning curves, you will get a better insight into your training, and you will be able to fix bias-variance problems to some extent. They also allow you to predict your errors for larger H and n by observing the shapes of curves.

> *During training, the bias and variance of the AI model should be estimated using learning curves or by measuring the training error and the generalization error. Overfitting and underfitting problems are related to bias and variance trade-offs and appropriate strategies should be employed to avoid these problems.*

5.3   From the Lab to the Dirty Field

Consider these scenarios:

1. We train an AI model to automatically diagnose retina images for retinopathy. Thanks to great work by our AI engineers, we are able to achieve very high accuracy. We decide to deploy it in the field and are able to convince hospitals and clinics in India to test it. When the results of the field trial pour in, we are aghast. The accuracy is far below what we expected. What went wrong? We find out that we trained on images acquired in the USA, which are different in subtle ways from those in India due to differences in quality and age of machines. Furthermore there is a wide variation in the data from one clinic to another and from one nurse to another within the same clinic which our training set did not possess.

2. We train an AI model to recognize speech. We train it on hundreds of hours of training data. But our customers are not very satisfied. When we perform a detailed analysis of the mistakes which the model is making, we realize that background noises show a far greater diversity around the world due to a variety of reasons.

3. Our customers are demanding a feature for document classification which we can provide using AI. However they are not willing to share their proprietary data. We train our model on documents, articles, and news stories obtained from the Internet. We integrate the solution in a new product offering after achieving high accuracy. We keep our fingers crossed while awaiting customer feedback, because we have an uncomfortable feeling that the accuracy will not be as high on their proprietary data.

The above problems are because of a mismatch between the training set and the field set. They are not sampled from the same underlying distribution, a departure from the preferred situation. A *training-test mismatch* is a common problem and shows the gap between lab settings and the real world. To solve the problem, an investment in resources to get actual samples from the field is needed. You should include these field samples in the training set and the development set. Experiments on modifications of the loss function such that field images are weighted more heavily can be conducted. If it is not possible to get more data, evaluate your training on a development set derived from the training set itself, ensuring that you have high accuracy and your system is working fine as expected. This will imply that the system is likely to become better when field samples become available. Gradually adapt the model to the new distribution using fine-tuning or transfer learning approaches.

5.4   System Design

Often a complete AI solution will involve many components, and you will face system design choices. One choice will be between component-based design and an alternative in which different components are integrated into one trainable architecture.

Sometimes it is a great idea to perform End-to-End (E2E) learning. That is, we input the raw data to an AI model, and magically out comes the final answer. We may have enough data to train such a powerful AI model. Even if there is an E2E solution, an investment in resources is needed to ensure that the data is properly annotated and preprocessed.

At other times, it is critical to break up the solution in an ingenious way. The required expressive power of an E2E solution may be very large, making it practically infeasible due to lack of enough training data. At the same time, robust and highly accurate solutions may already exist for some of the proposed components, which will facilitate a component-based design. For smaller components, the data requirements will be more manageable. Here are a few examples:

1. For Optical Character Recognition (OCR) in the wild, first detect the location of text, and then read the text in those detected text boxes.
2. For fine-grained classification, first detect the primary category such as vehicles, birds, and beetles, and then classify the image region.
3. From speech recognition to speech synthesis in a Question And Answer (QA) dialog with a home assistant device, first perform acoustic modeling, and then combine the phonetic output with other components such as linguistic features and a language model to transcribe the spoken sentence. Speech to text is an example in which high-performing E2E solutions have been built, and therefore both component-based design and integrated design are acceptable. The extracted sentence is then input to an NLU system which outputs the answer, which is then

sent to a speech synthesizer. It is hard to imagine an E2E waveform-to-waveform solution which integrates everything for the whole task.

4. In analyzing a street scene for traffic, social, and economic analysis, different components will first detect pedestrians, vehicles, buildings, roads, and traffic signals. Alternatively one can perform instance-aware holistic semantic segmentation. Fine-grained classification labels and different attributes of the detected objects are then computed by other components. At the end, aggregate statistics are produced.

Such a component-based design can facilitate debugging of the whole system by looking at the directed acyclic computation graph of different components. If any given component fails on perfect input, then it means it will fail on the output of the preceding components. You can easily check whether a component is contributing to the overall system error by evaluating the performance of other preceding components on whose output it is causally dependent. If the preceding components are working fine, then it means that the given component has problems which need to be fixed. In other words, the system is debugged by evaluating components closest to the input first and then moving to other components in a breadth-first traversal of the system.

One of the main differences between AI and classical ML is that the former works in the raw data space and performs automated feature engineering. This may not always be the best idea. AI can be assisted by providing some additional input or by enriching its input. This will help the training of AI models by not wasting their neurons on something which can be computed externally. This is especially useful if the raw data space is very large. Here are two examples:

1. For video data, provide the motion field (optical flow) as an additional layer in the input. See Chap. 9 for details.
2. For speech recognition, provide a spectrogram as the input rather than the raw waveform. See Chap. 11 for details.

We mention that for the second example, it is possible to use the raw waveform by making the AI model learn a spectrogram-like representation of the input; see Chap. 11.

Observe that enriching input as above is a special type of feature engineering where the dimensionality of the augmented features is of the same order as the input. In this sense, it is very different from feature engineering in classical ML, where there is significant reduction in the dimensionality.

> *The real world is different from laboratories, and therefore a conscious effort should be made to reflect the real-world distribution in the data being used in research. End-to-end learning is a powerful mechanism capable of solving several problems, yet at the same time, component-based design should be employed for complex systems. Furthermore, the raw data space can be enriched in order to assist the AI model.*

5.5 Flavors of Learning

Till now we have believed in the assumption that learning is all about solving a given problem, that there is one training dataset, and that there is one loss function. This is a great start. In this section, a few more interesting formulations are discussed.

The conventional goal of training an AI model is to create a specialist model. The AI model becomes the world expert in detecting cats. What if there is a new task? Suppose we take the cat-detecting AI model on an African safari and we want it to detect lions and zebras. A naive approach is to repeat the whole process of data collection and training. A key observation is that while getting trained to detect cats, the AI model built a robust hierarchical representation of features, and these features are similar to what one needs to detect the new categories of lions and zebras. The first few convolutional layers specialize in detecting edges, textures, and color blobs, and will readily transfer to the new task. The higher layers will likely need fine-tuning. Therefore if the cat AI model has L layers, we can freeze the first $L - k$ layers, for a small k, and learn the top k layers by running backpropagation through them and tweaking their parameters further. This is called *fine-tuning* or *transfer learning*, and it is a powerful approach which reduces the need for training data for new tasks. In practice, you download a pretrained model from the public domain and fine-tune it for your own task.

At times, the goal is to learn to learn. Here is an analogy. When you are in K to 12th grades, the goal is not to learn a highly specialized field but to reach a point from where you can learn other tasks. The emphasis is on building a strong foundation and on learning general skills. In the world of AI, the goal is to train an AI model toward a point from where it can learn new tasks. This is called *meta-learning*. During meta-learning, you show the AI model new example tasks and tweak your parameters in such a way that it will provide a good starting point for the new tasks. Therefore, there is a look-ahead when optimizing the starting point in which we look ahead at the future when new tasks are learned; see Fig. 5.5. In transfer learning, the starting point for new tasks is a pretrained model. In meta-

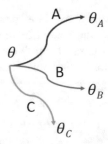

Fig. 5.5 Meta-learning seeks to find a good initialization of parameters from where multiple tasks can be learned by fine-tuning. In the figure, three tasks A, B, and C can be learned starting from the initial point θ. In order to find the best θ, we have to compute the gradient of the loss at the future points θ_A, θ_B, and θ_C with respect to θ

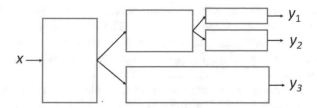

Fig. 5.6 In multi-task learning, several different outputs are produced for a given input. A large number of choices exist for designing and training a multi-tasking AI model. For each training input sample, not all of the output annotations may be available. Certain tasks may be easier than the others. Each task has its own loss function, and optimizing for one may affect the performance of the model on other tasks. The training data requirements for different tasks may be different. For self-driving cars, an optimal solution of the challenging problem of designing and training multi-tasking AI models is crucial

learning, one explicitly optimizes the initial model by fine-tuning it using new example tasks in the hope that for an unseen new task, the same fine-tuning will work; see [172]. In general, meta-learning optimizes parameters which control a subsequent learning process. In the above case, these parameters are those of the initial model.

Will you be happy just being able to do one task well? Being able to learn many related tasks is a worthy goal, and this is called *multi-task learning*. Based on the idea of transfer learning, there is an easy approach to implement multi-task learning. Layers of an AI model are used as the backbone architecture, and multiple heads for different tasks with their customized loss functions are attached to this backbone model. Multi-task learning is important for complex applications such as self-driving cars in which one has to perform multiple tasks. One needs to detect pedestrians, vehicles, traffic signs, traffic lights, road markings, and lanes. Instead of having separate AI models for these tasks, it is useful to perform multi-task learning because different tasks share common features and the same training data can be used. The design space of a multi-tasking AI model can be very challenging; see Fig. 5.6 for an example architecture, and see [203].

In *semi-supervised learning*, the idea is to use unlabeled data. One can first train a model using labeled data and then create pseudo-labels for unlabeled data using this model and include those samples with high confidence scores for a second round of training. For a sample without label, one can aim for consistency of predicted labels on its different perturbations as a regularization constraint. Note that we are making smoothness assumption about the carved manifold in order to make use of unlabeled data. Empirically, these ideas seem to help. In contrast, *self-supervised learning* first builds latent representation using unlabeled data and then fine-tunes the model for specific tasks using supervised learning. It has led to improvements in NLU, and it is an active area of research for visual data.

Suppose you are offering an online project-oriented course for gardening to students who are distributed around the world. You teach them how to grow a garden where plants and trees flourish. Students share their experiences, and it is collaborative learning in which each student develops his/her own model based on the local data. This is called *federated learning*. The distinction from the commonly used *distributed learning* is that each local model is adapted to its own local data. At the same time, there is sharing of models, which can potentially improve them. In distributed learning, data is homogeneous and there is one single model. In federated learning, data is heterogeneous and there are multiple models. An application of federated learning is in building customized models on individual mobile phones.

In AI, the problem of making an AI learn with a small training dataset is called *few-shot learning*, which is motivated by the observation that humans need only a few examples to learn. It is possible that our brains have pretrained networks, as a result of three billion years of evolution, which get fine-tuned using transfer learning with a few examples. This is a challenging area of research and holds the key to further advances in AI.

Finally, we describe an interesting design technique which could help in some applications. We consider the situation in which the input space can be divided into clusters or regimes, for each of which a separate expert network can be trained. We consider the example of semantic segmentation of cats in which the goal is to produce the contours of a given cat. Suppose the cat manifold has two well-separated subsets, one for fluffy cats and the other for non-fluffy ones, and then we can have a mixture of two expert CNNs for semantic segmentation. A learnable gating network will hopefully learn to decide the type of a given cat, and the corresponding expert will be invoked. This is similar to component-based design except that here we don't have ground-truth labels for fluffy cats, and therefore we cannot explicitly train a classifier. During backpropagation, the gating network learns to divide the input space into two clusters. These two clusters will hopefully correspond to fluffy and non-fluffy cats, though the result really depends on the geometry of the input space and what exactly the gating network carves out.

A *mixture-of-experts* (MOE) layer consists of N experts $E_1, \ldots, E_N$ and a gating network G whose output is an N-dimensional vector. For a given input x, the output of the MOE layer is given by

$$y = \sum_{i=1}^{N} G(x)_i E_i(x).$$

If the output of the gating network is a sparse vector, then this conditional computation, which invokes parts of the network on a per-example basis, leads to a saving in computation and a better utilization of the capacity of the network; see [192]. The experts can share common layers which compute features useful to all experts. It is crucial to add noise to the outputs of the experts during training in order to strike a balance between exploration and exploitation; otherwise those experts which are initially lagging behind may never get a chance to learn properly.

5.6 Ingenuity and Big Data in the Success of AI

Finally, we make a comment to understand further why AI has been successful. In Chap. 4, we learned that there are two major reasons behind the success of AI: (1) its high expressive power, that is, high carving capacity to approximate manifolds and functions and (2) the SGD algorithm which is unlikely to get trapped in a local minimum. Unlike classical ML, AI doesn't underfit for challenging problems in image, speech, and language understanding. This chapter adds one more piece to explain why it doesn't overfit either. If there is an innovative design of an AI model for a specific problem, say for image understanding, then this reduces variance. We learned that the learning curves show that a high-capacity model with adequate training data will generalize very well. An ingenious design means that we are looking at the underlying manifold structure to design an efficient way to carve it. The convergence of brilliant research and an explosion of data have led to the AI revolution.

> *Besides the reason of the very high capacity of AI models converging to near-optimal solutions in the optimization landscape, which explains why they don't underfit to the training data for challenging problems in AI, there is a reason which explains why they don't overfit. Because of the innovative designs of AI models which reduce variance, AI models with exponential expressive power generalize well to unseen data, when trained on an adequate amount of training data.*

5.7 Chapter Summary

The training of an AI model is a process which has to be governed carefully. The starting point is the organization of data in three categories of training set, development set, and test set. The results of training should be evaluated on the development set and later on the test set. A variety of tools such as the ROC and PR curves can assist in the evaluation. At times, adding a regularization term helps the training process by reducing overfitting. The concepts of bias and variance should be employed to characterize the specific training situation in terms of overfitting and underfitting. Learning surfaces and learning curves are helpful in the visualization of errors, and bias-variance trade-off, and they provide a quick insight into the performance of the AI model. Ideally the AI model should have low bias and low variance, which is achieved with a careful design of the system. Trained AI models may not work well when deployed in the real world because the statistical distribution of data in the field may be different from the training data distribution in research laboratories. For complex problems, it may be easier to break up the system into components instead of building an E2E solution. Though AI works great in the raw data space, by enriching the raw data input, it is possible to guide the AI model toward an effective solution which minimizes the need for data. Besides task-specific learning with a single training dataset, there are other forms of learning. In particular, transfer learning, meta-learning, multi-task learning, federated learning, and few-shot learning have been attracting the attention of the AI community. Finally, the ingenious design of AI models combined with large training data ensures that they generalize well to unseen data.

Problem Set 5

1. Why is the ROC curve monotonic? In general, why is the PR curve not monotonic?
2. Provide a geometrical intuition about the Lagrange multiplier method for constrained optimization.
3. What do the following numbers tell you about the bias and the variance? Do they indicate a problem which needs to be solved? How will you solve it?

    (a) 20% training error and 21% test error.
    (b) 1% training error and 21% test error.
    (c) 20% training error and 40% test error.

4. What insights do learning curves provide?
5. What could be a potential problem in the training of a mixture of experts?
6. Why multi-task learning is important for autonomous vehicles?
7. Where will you not use multi-task learning?

Chapter 6
Unleashing the Power of Generation

> It is only by drawing often, drawing everything, drawing incessantly, that one fine day you discover, to your surprise, that you have rendered something in its true character.
>
> Camille Pissarro

> Those who do not want to imitate anything, produce nothing.
>
> Salvador Dali

At the highest level, an AI model is a mathematical function. It is our interpretation of the input and the output which makes the function of interest to us. When we hear the term AI, we often think of data samples as the input. In generative AI, the goal is to have data samples as the output.

If you are asked to paint a flower, you may ask many questions. "What type is it? How many petals? How many sepals? What is its size? What should be the color? Should it be bending at an angle?" What you are doing is that you are collecting the generative parameters of the flower. From these parameters, you will create a painting.

AI does the same with one difference. You knew the important parameters for flowers in advance. In AI, we don't have training datasets in which there are so many annotations available. Furthermore, not all parameters can be known exhaustively even if we were to invest in data annotation. Often AI is just handed a bunch of classes, with the class label being the only annotation. Therefore it works with latent parameters rather than known parameters. This makes the problem challenging. In this chapter, we will use the notions of known interpretable parameters and unknown latent parameters to build intuition into generative AI.

6.1   Creating Universes

In supervised machine learning, we work with training data, which in general can be both real collected examples and synthetically generated examples:

$$D_{\text{training}} = D_{\text{real}} \cup D_{\text{generated}}.$$

In the cosmological constants exercise below, the training data is all synthetic, which is an application area of generative models. There is only one real data sample, which is the universe we are living in, and which is the test set. For recognizing whether there is a cat in a picture, all data is real as people love to upload pictures of their cats on the Internet, and there is no shortage of data. For self-driving cars, we use simulated data in addition to the real-world data.

Any computable function is a generative function if it generates objects of interest. Being interesting is a subjective and anthropocentric attribute, and there are myriad types of creation. An object can be an image, video, animation, music, speech, text, data, simulations, design, blueprint, etc.

What could be more fun than creating whole universes?

The universe we live in with its scientific laws and cosmological constants is the most sophisticated generative model. Thanks to this model, we are here. Here is an exercise which allows us to generate universes using mathematics and physics, which also illustrates the power of generative methods and AI; see Fig. 6.1.

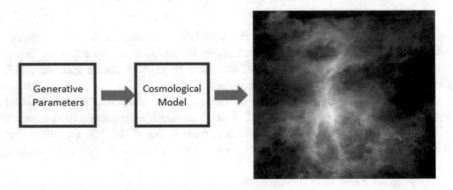

Fig. 6.1 A cosmological model is a generative model. (The Orion nebula image: courtesy ESA/NASA/JPL-Caltech using Herschel Space Observatory)

*** **Exercise 16   Cosmology and AI**

Based on a literature survey, discuss how generative models can be used to estimate cosmological constants. Is it possible to speed up these methods using supervised training methods?

6.2 To Recognize It, Learn to Draw It

We use our generative powers not only to create and pursue high-level intellectual activities but also to make sense of what we perceive in daily life; see Fig. 6.2.

At the most fundamental level, how do you recognize a cat? Children can draw sketches of cats, and those with artistic abilities can make beautiful paintings. There is something in our generative model we must be using to assist image recognition. There are two possibilities for cats:

Brute Force   Generate all possible cats and see whether the observed cat belongs to that set. This is definitely not the way we recognize cats.

Simplex method   Start with some generative parameters and its corresponding cat. Tweak those parameters in such a way that the generated cat transforms into the cat you are observing. If you can't transform, then it is not a cat. The simplex-tweaking method resembles the working of a forensic artist who can draw a sketch of a person with the help of an eyewitness who provides generative parameters which get gradually tweaked till the sketch becomes closer to the real face.

The second explanation seems more reasonable. Since we recognize cats in a fraction of a second and most of the processing is unconscious, we don't really

Fig. 6.2 Humans have a remarkable generative model, and we can conjecture that it connects with our robust perception abilities. Similarly, it seems that the line between discriminative AI and generative AI is not hard. (Credit for the image, c. 1896: artist Camille Pissarro (1830–1903), courtesy of the National Gallery of Art, Washington DC)

know whether this is really what happens. A child sees a black cat and abstracts out black color as a generative parameter. The same child sees a white cat and is able to tweak the original example by changing its color. A child sees a cat turned left. Then, it sees a cat turned right. The child is able to transform the pose of the first into that of the second. One training example will suffice to recognize all cats. It is often said that AI models need lots of data whereas humans can learn from a single example. This may change with research in generative AI which in turn leads to a robust solution to self-supervised learning and therefore to a potential breakthrough in AI, allowing AI to learn from unlabeled data. Discriminative AI and generative AI may turn out to be two sides of the same coin. In fact, generative methods to build latent representations under the reconstruction loss followed by problem-specific supervised fine-tuning have been inspired by this connection [90].

6.3 General Definition

See Sect. 1.5 for an introduction to generative models. Let X be a set of objects of interest. Recall that a *computable generative model* is a Turing computable function g such that its range is the set X and its domain is the set of valid generative parameters Θ. If we endow Θ with a probability distribution P, then we have a probability distribution over X:

$$g : \Theta \to X,$$

$$P(x) = \sum_{g(\theta)=x} P(\theta),$$

which can be generalized to K multiple classes,

$$g : \Theta \to X_1 \cup \ldots \cup X_K.$$

Often generative models are defined as those from which one can draw data samples as per its probability distribution. In other words, a discriminative model is defined as one which models the conditional probability $P(y|x)$, and a generative model models the joint distribution $P(x, y)$. The above definition subsumes the conventional definition.

** Exercise 17   Sampling

Given a computable generative model g endowed with a probability distribution P over its range X as defined above, specify how one can randomly draw samples from X as given by P.

6.4   Generative Parameters

Let's say you are building a computer graphics software package, which is an implementation of a Turing computable function g, which generates pictures of cats. A list of input parameters is

$$\{\theta_1, \theta_2, \ldots, \theta_p\},$$

where the parameters are real numbers describing physiology, color, texture, size, orientation, pose, and all the attributes which define what it means to look like a cat. Assume that the software is working very well and it renders realistic-looking cats and dogs. As the numerical parameters are varied continuously, we expect the images to change continuously. Let the resolution of the images be $n = w \times h$ pixels. Consider the mapping

$$g : \Theta \to \mathbb{R}^n,$$

where $\Theta \subseteq \mathbb{R}^p$ is the set of valid generative parameters (attributes). Its range is the cat manifold. The same idea applies to speech and audio. From the shape parameters of a drum or the physiological parameters of the vocal chords of a singer, we can generate the manifolds of beats and of vocal sounds, respectively, and small variations of these generative parameters will result in small variations in the output waveforms.

Let us build a notion of Kolmogorov or algorithmic complexity of cats. The sum

$$|T_g| + p,$$

of the size of the smallest Turing machine T_g implementing g and the size p of the parameter vector, measures the "complexity" of a cat. We expect $p \ll n$ though $|T_g|$ may be large. This is the basis of the manifold learning hypothesis, which states that a class manifold $g(\Theta)$ is a lower-dimensional manifold embedded in a larger space $\mathbb{R}^n$. As generative parameters vary, g creates a p-dimensional cat manifold embedded inside $\mathbb{R}^n$.

In AI, the Turing machine is approximated by an AI model using data-driven methods and without knowing what p and Θ are.

> *Generative models play an important role in creating simulated and synthetic data. Generative parameters (features) form a latent semantic space representing the most important features of an object. Having a robust generative model and the ability to modify the generative parameters to create a given image will provide a solution to the image recognition problem and will be an alternative to discriminative AI.*

6.5   Generative AI Models

According to our definition, a Turing computable generative map exists:

$$g : \Theta \rightarrow \text{Cat Manifold},$$

where $\Theta \subseteq \mathbb{R}^p$ for an unknown p. Unfortunately, this perfect model is inaccessible to us, and it is not a supervised problem in the conventional sense. We don't have input-output pairs. Therefore generative AI asks:

1. What is an approximate range of p?
2. Given training images of cats, can we approximate g by an AI model:

$$F : \Theta' \rightarrow \mathbb{R}^n,$$

   for some $\Theta' \subseteq \mathbb{R}^q$ and some q such that its range is approximately the cat manifold? That is,

$$F(\Theta') \approx \text{Cat Manifold}.$$

An answer to the first question can be found only by trial and error. We just evaluate different values of q in the second question. We will look at some options for F in the second question in the next section.

Since we don't have access to the actual Θ, and F is an approximation of g, therefore Θ' is called the latent variable space. We use generative variables, generative attributes, and generative parameters interchangeably in this book. These could be either latent or known.

One can compute an upper bound on the Kolmogorov complexity of cats by

$$|M| + q,$$

where $|M|$ is the size of the smallest AI model M generating cats and q is the size of its input.

6.5.1   Restricted Boltzmann Machines

Restricted Boltzmann machines (RBMs) have a rich history in the study of neural networks and are inspired by physics. There are concepts of configuration and energy as in physical systems consisting of interconnected units such as spin glasses and Ising models. The lower the energy, the higher the probability. An RBM is a special case of graph-based Boltzmann machine; see [85]. A stacked Restricted Boltzmann Machine (RBM) is called a Deep Belief Network (DBN); see the influential paper [90]. DBNs can also be constructed by stacking autoencoders, which are discussed in the next subsection. We devote this section to RBMs and DBNs because of their role in revitalizing the field of neural networks in 2006.

In an RBM, we have a set of visible neurons (data neurons) v, which are connected to hidden neurons (feature detectors) h representing generative parameters. They form an undirected bipartite graph. Each neuron is a linear regression module with weights and bias term and the sigmoid activation.

The commonly studied case is that of binary neurons. The activation values of the neurons are the probabilities of a neuron having value 1, and an RBM defines a probability distribution from which we can stochastically sample binary data and binary features. Given a binary h, visible neurons get activated, and the output of the neuron v_i is $\Pr(v_i = 1|h)$. Similarly for h, given a binary v. The probability space is sampled using Gibbs sampling, which alternates between conditional probabilities $\Pr(h|v)$ and $\Pr(v|h)$. Therefore one can map a vector from the latent variable space to the data space and vice versa. Note that an RBM forms a special encoder-decoder architecture:

$$h = E(v),$$
$$v' = D(h),$$

where E and D share weights since the edges are undirected in the bipartite graph.

A grayscale image of a cat with pixel values in the range $[0, 1]$ is interpreted as probabilities, and it can be binarized stochastically. A binary cat v is sampled from it. The conditional $\Pr(h|v)$ is computed, which in turn can be stochastically binarized. The binarized h generates $\Pr(v|h)$ and we are back to the image space. Continue the process for one more step. These steps allow one to compute the gradient steps to maximize the log-likelihood of data and to tweak learnable parameters; see [88, 90]. The Kullback-Leibler divergence between the data distribution and the equilibrium distribution defined by the model gets minimized. The algorithm is known as single-step contrastive divergence.

One can treat the hidden units of the trained RBM as visible units for a second RBM which is stacked over the first one. After layer-by-layer greedy training of several layers, the entire model can be trained as one big autoencoder by unfolding the decoding layers of the RBMs on top of their stacked encoding layers, and at this time the weights in the encoder and decoder networks need not stay the same:

$$z = E_3(E_2(E_1(v))),$$
$$v' = D_1(D_2(D_3(z))).$$

For a classification or regression task given by input-output pairs (x, y), one can take the output z of the encoder, which forms a low-dimensional latent representation of the input, and attach another network F to it and fine-tune the whole DBN network under supervised loss:

$$z = E_3(E_2(E_1(x))),$$

$$y = F(z).$$

There is an approach called neural autoregressive distribution estimation; see [217], inspired by RBM, in which the probability product rule is utilized to yield a tractable estimator of probabilities. For D-dimensional data, one uses FNNs to model D many one-dimensional, conditional probabilities. For example, for $D = 3$ binary data, we write:

$$P(x_1, x_2, x_3) = P(x_1)P(x_2|x_1)P|(x_3|x_1, x_2),$$

and use three FFNs to model the three conditionals. In general, use an FFN to compute

$$P(x_k = 1|x_1, x_2, \ldots, x_{k-1}).$$

There is a quadratic increase in the number of model parameters in terms of D, but it can be kept in check with weight sharing.

6.5.2 Autoencoders

An autoencoder has an encoder-decoder architecture:

$$\theta = E(X),$$

$$X = D(\theta),$$

where E is the encoder, D is the decoder, and θ is interpreted as the (latent) generative parameter vector. The encoder and the decoder are trained jointly under the reconstruction loss:

$$d(X, D(E(X))),$$

where d is the L^2 distance. See Fig. 6.3. If $|\theta| < |X|$, then this gives PCA-like dimensionality reduction. The encoder-decoder architecture for sequences is a variation of the autoencoder; see Fig. 3.5.

A variation of autoencoders exists to deal with situations when we want $|\theta|$ to be large, that is, $|\theta| \geq |X|$, in which case we want to discourage the network from learning the identity mapping. This is achieved by regularization. In a sparse autoencoder, the parameter vector is constrained to be as sparse as possible. A sparsity-encouraging term is added to the loss function as a regularization term, making use of the Lagrange multiplier method for optimization with a constraint:

$$\text{Loss Function} = (\text{Reconstruction Loss}) + \lambda^* (\text{Sparsity Loss}).$$

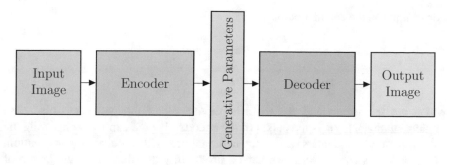

Fig. 6.3 Encoder-decoder architecture of an autoencoder. The encoder encodes a representation of the input, which is decoded by the decoder back into the input. The network is trained under the reconstruction loss. The output need not be the reconstructed input, and in general it could be any desired output, and then the architecture can be viewed as a translator. The architecture can be generalized to sequential data. In fact, the reader will notice while reading the book that several solutions in NLU, speech, and computer vision can be conceptualized in the form of encoder-decoder architecture. This connects with the observation that AI models build distributed representations of the input

The L^1-norm penalty can be used for the sparsity loss and is provided by the AI software framework in the form of an activity regularizer. Therefore, we have the regularized loss as

$$\|X - D(E(X))\|_2 + \lambda\|\theta\|_1.$$

Another variation of the autoencoder is the denoising autoencoder. Random noise is added to the input image X in order to slightly corrupt it. The autoencoder learns to ignore the noise ϵ in the input $X + \epsilon$ and create robust generative parameters. See [219] for background, and see [59] for the L^1 sparsity regularizer.

6.5.3 Variational Autoencoder

The problem with autoencoders is that we don't control the statistical distribution of the latent generative parameters θ. We don't know where the cat manifold maps back into the generative space, and therefore we don't know the valid values of the parameters. We want the mapping from the generative latent space to the image space to be smooth.

A Variational Autoencoder (VAE) is a generative model in which the encoder outputs statistical parameters of the parameters for more effective control. Suppose the latent generative parameter vector is

$$\theta = \{\theta_1, \theta_2, \ldots, \theta_p\}.$$

For an input x, the encoder outputs

$$\{\mu_1, \mu_2, \ldots, \mu_p\},$$

$$\{\sigma_1, \sigma_2, \ldots, \sigma_p\},$$

which are the means and the standard deviations of the parameters. In some implementations, $\log \sigma$ is the output of the encoder. We are explicitly endowing the valid generative parameters which map to the input x with a probability distribution. Therefore for a cat image x, we have a probability distribution on the length of its tail, with a mean length and an associated variance. The decoder randomly draws a sample from the Gaussian distributions with these means and variances to reconstruct the input x. This approach makes the mapping of the latent parameter space to the data space continuous, making the generative parameters interpretable. Note that a VAE performs *probabilistic* mapping unlike an autoencoder which is deterministic. For a given cat image, we have multiple reconstructions. This explicitly encourages smoothness of the generative mapping. Nearby latent vectors map to similar reconstructions. After the training session, we can sample with confidence from the hyper-spherical vicinity of latent vectors for the training samples, leading to visually better results.

There is a problem with this approach, which is that sampling in the middle of a neural network can't be differentiated. Fortunately, the problem is solved using a technique known as the reparameterization trick. For each parameter, a standard Gaussian distribution with mean 0 and variance 1 is sampled and fed to an auxiliary input layer, which multiplies it by the standard deviation σ and adds the mean μ, which are output by the decoder. In other words, the architecture consists of an encoder E and a decoder D, and the following steps:

$$y = \text{Sample}(\mathcal{N}(0, 1)),$$

$$\mu, \sigma = E(X),$$

$$\theta = y * \sigma + \mu,$$

$$X = D(\theta).$$

Implemented as above, the encoder may start learning $\sigma = 0$, reducing it to the standard deterministic autoencoder. We want the mapping to be probabilistic. Therefore, in addition to the reconstruction loss, the Kullback-Leibler divergence term is added to the cost function as a regularization term to penalize the departure of the output of the encoder from the standard Gaussian distribution:

$$\text{KL}(\mathcal{N}(\mu, \sigma)\|\mathcal{N}(0, 1)) = -1/2(1 + \log \sigma^2 - \mu^2 - \sigma^2).$$

This will encourage the encoder to not output 0 for the variance. The total loss function is

$$\text{Loss Function} = (\text{Reconstruction Loss}) + \lambda^* (\text{KL Loss}).$$

A fine-tuned value of the Lagrangian multiplier λ is critical to balance the reconstruction loss with the goal of learning interpretable latent factors. If the encoder is still learning small values of the variances, then λ can be increased.

See [31, 82, 112] to learn about VAE. The idea is based on variational inference in statistics, which can be used for approximate inference of probability distributions. This is in contrast to the MCMC algorithm, which gives exact inference at a higher computation cost.

6.5.4 Pixel Recursive Models

Consider the approach of creating an image pixel by pixel, row by row. One can use the context of the previously generated pixels to model the probability distribution of the next pixel, and sample from it. This is similar to language modeling using multilayer LSTM, in which one predicts the probability of the next word in a sentence. Therefore it is pixel modeling with the additional modification of using the 2-D geometry of an image in which dependence flows in two spatial directions. An image has become like a piece of 2-D text where pixels are the words.

The receptive field for a pixel is unbounded and requires sequential processing in this autoregressive approach during training. To parallelize the processing during training, one uses a bounded receptive field and uses convolutions on the fixed size, bounded neighboring contexts. At the inference time to generate an image, the processing is still pixel by pixel. See [155, 184] for this autoregressive approach known as Pixel RNN. A similar idea of using the previous context to predict a pixel has been used for data compression in the past, where the prediction was based on hand-engineered rules. The idea can be applied to generate raw audio waveforms by using causal convolutional filters which are dilated to increase the causal context of the previously generated samples; see [156].

6.5.5 Generative Adversarial Networks

It has been suggested in the Matrix movie (1999) that the whole world is the Matrix and we are in a simulation. Alas, there is no way for us to discriminate the real from the virtual. We keep on living in this world thinking it is real. The AI which runs the Matrix could have adopted an approach of trial and test: generate a simulation and see whether people suspect that it is fake, and keep iterating till they take it for real.

Generative Adversarial Network (GAN) takes this approach based on discriminating the real from the fake. For details see [61]. Instead of the reconstruction loss used in autoencoders, the generative model in GAN is evaluated by an adversary, which is a classifier trained to distinguish real, training images from fake, generated images. The generator seeks to fool the discriminator. The discriminator does it best to keep from being fooled. GANs can be understood in a game-theoretic framework where two players seek a *Nash equilibrium* of a game. The Nash equilibrium of the game is reached when the generator can synthesize images which are indistinguishable from real images.

The generator G and the discriminator D get trained in alternative epochs. The output of D is evaluated to compute the loss. In the discriminator training, G is kept fixed and backpropagation is only through D, whose weights are updated. In the generator training, backpropagation runs through D to G, and only the learnable weights of G are changed. For the discriminator training, the loss function is the binary real-versus-fake cross-entropy loss:

$$-\mathbb{E}_{x}[\log D(x)] - \mathbb{E}_{\theta}[\log(1 - D(G(\theta)))],$$

where $\mathbb{E}_{x}$ is the expectation over the training set images and $\mathbb{E}_{\theta}$ is the expectation over the generative parameters which are input to G. For the generator training, the loss is in the other direction and is simpler since the first term $\mathbb{E}_{x}$ is unaffected by G:

$$\mathbb{E}_{\theta}[\log(1 - D(G(\theta)))].$$

A more stable alternative to the generator loss is

$$-\mathbb{E}_{\theta}[\log D(G(\theta))].$$

Adversarial GAN can be generalized to autoencoder GAN. It is a natural idea to combine variational autoencoders with GAN; see [181] for further reading. The adversarial loss is combined with the reconstruction loss.

6.5.6 *Wasserstein Generative Adversarial Networks*

One problem which is encountered during training of GANs is that of mode collapse. For example, a cat GAN can just keep on generating pictures of a few cats. In addition to mode collapse, the training of a GAN can suffer from vanishing gradients when the discriminator is doing its job too well and not giving an opportunity to the generator to learn. A third problem has been the failure of training to converge.

This relates to the bigger question of how to evaluate any generative model. For example, how will mode collapse be detected? How can we measure the progress of research in generative AI? Current ideas for evaluation include checking whether different classes have similar ratios of numbers of generated samples as real classes and whether the features of real and generated images computed by different layers of a pre-trained classifier have similar statistical distributions as measured by a suitable statistical distance, such as Fréchet inception distance [81]. Moreover, the generated samples should be novel, that is, their nearest neighbors obtained by running an image similarity algorithm on the real dataset should look different.

The original GAN has been generalized to several different kinds over time in order to improve the training. In this section, we describe Wasserstein GANs, which are easier to train than the original GANs. The Wasserstein loss is derived from the Wasserstein distance, also known as the earth mover's distance, which compares two probability distributions to compute the loss. The distance is based on the optimal transport theory in which one probability distribution is morphed into the other by transporting probability mass. The output of the discriminator does not have to be a probability between 0 and 1. It can be an arbitrary real number, and there is no need to squash the output. However, this flexibility means that the weights of the networks are required to be clipped so that they remain restricted to a range in order to constrain the output to a stable range.

Let us develop further intuition about the Wasserstein GAN. Suppose you want to generate cats using a GAN. Initially, the generator will generate very random-looking images, and the discriminator will quickly learn to distinguish them from the real images. The discriminator loss will rapidly reduce and the discriminator will approach the minimal loss. We will say that it is working near its optimal point. At the same time, the generator will have a hard time learning to generate cat images. This happens because the loss for the generator has saturated and its gradients are very small. The fake images and the real images are very different, and their respective manifolds are far apart and disjoint. Mathematically, we can state the vanishing gradient problem as

$$\lim_{D \to D^*} \|\nabla_\theta (\text{Generator Loss})\|_2 = 0,$$

where D is the discriminator, D^* is the optimal discriminator, and θ is the set of learnable parameters of the generator.

This is like wanting to sample in San Francisco while you are far away in Atlanta. The gradients are very small, and they don't guide you how to move from Atlanta to San Francisco. This happens because we are dealing with probabilities which are bounded in the range [0,1] and they have become saturated. Therefore we want a loss function which is well-behaved on highly skewed probability distributions. In a way, this is reminiscent of the saturation of the bounded sigmoid function and how the unbounded ReLU speeds up training. We should experiment with other loss functions which have larger gradients when the fake images are getting very low probability scores. It should not be the case that we manage to initially make the gradients larger, but as we move to San Francisco, the training slows down considerably. Moreover, the gradient updates should be stable and not noisy.

What if we no longer squash the output into the range [0,1] and we drop the logarithm and just directly work with the output of the network as something to be optimized? In fact, this intuition works well. There are no vanishing gradients and there is no saturation of the loss. The loss steadily declines as we move from Atlanta to San Francisco. It is a well-behaved loss function. With this simple intuitive modification, moving westward from Atlanta does lead to a steady reduction in the loss. However, there is a downside to the unbounded output, and the training can become unstable with weights increasing without any bound. A practical remedy is to clip the weights in a finite range. One can formally work out the mathematics and show that now we are working with the Wasserstein distance. This earth mover metric is the least work needed to move infinitely fine sand of one probability distribution to another. The work done is defined as the mass of the sand multiplied by the distance by which it is transported.

In order to implement the Wasserstein loss, the discriminator no longer outputs a probability. As we pointed out, it has to output a real-valued number which is no longer bounded in the range [0,1]. The best is to interpret it as a numerical rating of the "realness" of an image. The discriminator is now referred to as a critic. The critic strives to assign a high rating value to the real images and a low rating value to the fake images. The generator strives to generate such images that the critic will rate them high. The Wasserstein loss function for the discriminator (critic) is

$$-\mathbb{E}_{x}[D(x)] + \mathbb{E}_{\theta}[D(G(\theta))].$$

For the generator, the Wasserstein loss is

$$-\mathbb{E}_{\theta}[D(G(\theta))].$$

Note that the loss functions are surprisingly simple. *We have changed probabilities to real numbers and dropped the logarithm.* With this simple change, training becomes stable and leads to better solutions due to effective handling of the vanishing gradient problem. See Fig. 6.4 and [4].

** Exercise 18   Wasserstein GAN

See [4]. Why does the use of Wasserstein loss lead to better training of a GAN?

6.6   The Carving Process in Generative AI

In the first chapter and Chap. 4, an intuition was developed into the fantastic geometrical carving process of manifolds and of functions for discriminative AI. The same conceptual understanding can be built for generative AI. We will build an intuition into this process which allows us to look at the working of AI models from a fresh perspective.

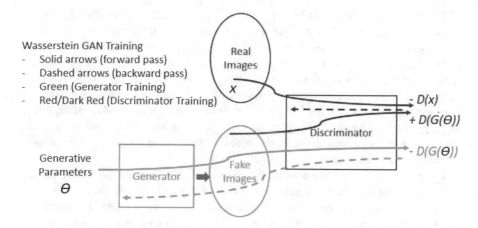

Fig. 6.4 Wasserstein GAN. In the generator training, though backpropagation runs through the discriminator, only the weights of the generator are updated. In the discriminator training, the generator is frozen, and the weights of the discriminator are fine-tuned for the real-versus-fake classification task

Consider generation of images. First note that generative AI for computer vision is a function fitting from the space of latent variables (generative parameters) to image pixels. Therefore the same function approximation process is working underneath as in discriminative AI, with some important differences.

Consider a simple example of creating binary images of vertical line segments. Suppose three *known* generative parameters are (t_x, t_y, l), which describe a binary image I of a vertical line of length l with its top end-point at location (t_x, t_y). Consider an image pixel at location (a, b). Consider the set of generative parameters for which $I(a, b) = 1$:

$$X = \{(t_x, t_y, l) | G(t_x, t_y, l) = 1\},$$

where G is the AI model with one output neuron generating the pixel value at the location (a, b). What is the shape of X? It is a set of points in a 3-D parameter space satisfying the constraints:

$$t_x = a,$$

$$t_y - l \leq b,$$

$$t_y \geq b.$$

The output neuron for the pixel (a, b) carves out this region in the parameter space. In general, X will be approximated by many convex polytopes whose number grows exponentially with the depth of the AI model.

For the above example of binary vertical line segments, verify that X is a 2-D convex polytope in the 3-D parameter space. For the neighboring pixel $(a + 1, b)$, the polytope just shifts along the t_x-axis. For the pixel $(a, b + 1)$, the polytope shifts in the (t_y, l) plane. The geometry of regions carved out by the output neurons is similar, and they can share the carving done by lower-level neurons. If the carving is accurate, then given the generative parameters (t_x, t_y, l), we will obtain the desired line segment. This is like the Hough transform in classical computer vision, which maps points in the parameter space to lines in the image space, and lines in the parameter space to points in the image space. In the Hough transform, the mapping is fixed. In AI, it is learned.

In this example, we have a semantic interpretation of known generative parameters which are not latent. In general, we don't have an interpretation, and we just estimate how many parameters are needed and consider them latent. Therefore it is not a supervised problem in the conventional sense. We are not given the ground truth in the form of (θ, I) pairs where θ is the generative parameter vector and I is the image.

In a GAN, we have a random vector as the input. In the data distribution of the random input vectors, some set of convex polytopes get carved out for each pixel (a, b) and fitted with a piecewise linear function. In autoencoders, by casting the problem as a supervised regression problem, we build a latent representation with a desired distribution from source images, which will get carved out for each pixel.

Interestingly, many of these learned latent parameters have an effect on the generated images in an interpretable manner. For the vertical line segment problem, if the AI model discovers parameters which relate to the position and the size of the segment, then it is learning something which is directly interpretable.

There is a wide selection of approaches to build generative AI models. Several of them are based on the reconstruction loss. GANs provide an alternative method to train a generative model by employing a discriminative model as an adversary. The performance of generative AI is becoming impressive with time, and it can generate highly realistic content; see Chap. 12. Generative AI shares the same fundamentals of the geometrical sculpting process as discriminative AI. In generative AI, the carving is done in the generative parameter space.

6.7   Representation of Individual Signals

So far we have been working under an assumption that we want to generate samples from a class of images, such as cats. We consider a different problem of generating pixel values of a given image of a cat. That is, we want to fit a pixel-generating function for an individual signal:

$$f : \mathbb{R}^2 \to \mathbb{R}^3, f(x) = y,$$

where y is the RGB value of the pixel at coordinates x. For 3-D shapes, we may want to fit a function:

$$f : \mathbb{R}^3 \to \{0, 1\}, f(x) = y,$$

where y is 1 if and only if the 3-D point at coordinates x is inside a given shape. In the same way, audio waveforms can be represented.

Two benefits can be immediately inferred from the neural implicit representation of signals and shapes. The first is data compression when the representation is more compact than the signal. The second is resolution independence, that is, the ability to render the signal at any desired resolution. The reader will notice that this is related to the conventional Fourier or wavelet representations of signals. The difference is that with deep neural networks, we are building a learned representation. Interestingly, if one uses the sine function $\sin(x)$ as the activation function instead of the ReLU function, one gets a better reconstruction with fine details, which is expected because we want the approximation to be not piecewise linear but smooth. See [201] for this new application of AI in representing individual signals which uses the sine function.

6.8 Chapter Summary

Human intelligence derives its robustness from its ability to generate objects of interest. Generative AI is a new frontier in AI research, and its first application is generation of synthetic training data. The process of generation involves representation of objects of interest in a semantic space of latent generative parameters (attributes). In generative AI, several methods have been experimented with, starting with RBM and autoencoders. Autoencoders are trained under the reconstruction loss and have several variations such as sparse, denoising, and variational autoencoders. Pixel modeling is intuitively the same as language modeling in NLU, where pixels are words in a 2-D document of an image. GANs formulate the problem using a novel approach in which generative parameters based on a collection of training images are learned in such a way that an adversarial discriminator can't distinguish the generated images from the real images. Wasserstein simplifies the loss function of a GAN in order to make the training stable. Generative AI uses the same geometrical sculpting process in the generative parameter space which is discussed in Chap. 4. Besides generating samples, AI can be successfully applied to the representation of an individual sample. For example, an image or a shape can be represented by an AI model, and this yields a resolution-independent compact representation.

Problem Set 6

1. How can generative AI potentially provide a solution to discriminative AI?
2. What type of learning is an autoencoder following?
3. Consider the following autoencoder networks in which the input is used to compute the L^2 reconstruction loss. In each case, assume that you are working with scalars. Suppose you run SGD algorithm to train the networks. To what values will the weight parameters converge?

    (a)

    $$y = w_1x_1 + w_2x_2,$$

    $$y_1 = y,$$

    $$y_2 = y,$$

    $$L = (y_1 - x_1)^2 + (y_2 - x_2)^2.$$

    (b)

    $$r = w_1x_1 + w_2x_2,$$

    $$y_1 = w_3r,$$

    $$y_2 = w_4r,$$

    $$L = (y_1 - x_1)^2 + (y_2 - x_2)^2.$$

    (c)

    $$r = w_1x_1 + w_2x_2,$$

    $$s = w_3r,$$

    $$y = w_4s,$$

    $$y_1 = y,$$

    $$y_2 = y,$$

    $$L = (y_1 - x_1)^2 + (y_2 - x_2)^2.$$

4. What can be done to ensure that an autoencoder doesn't just learn to copy the input to the output?
5. What property do VAEs try to encourage in the latent space?

6. Consider two probability distributions $P = [0.5, 0.5]$ and $Q = [0.1, 0.9]$. Compute the Kullback-Leibler divergence. Note that the KL divergence is not symmetric; therefore work out both directions.

7. Suppose a discrete probability distribution P has mass 0.5 at point (0, 0) and mass 0.5 at point (1, 1). Another distribution Q has mass 0.4 at point (0.5, 1) and mass 0.6 at point (0.5, 0.5). Compute the Wasserstein distance between the two distributions.

Chapter 7
The Road Most Rewarded

> Two roads diverged in a wood, and I –
> I took the one less traveled by,
> And that has made all the difference.
>
> Robert Frost

We all face critical decision-making stages in life. What major should I study? Who should I marry? What career should I pursue? Should I quit my job and launch a startup?

A 17-year-old high-school senior is in a special state. His/her action now will determine a new state in a few months. Which action to take? Parents, counselors, and teachers are eager to advise. They ask kids to evaluate expected long-term rewards. They encourage students to score high in tests and to go after the best, as it is the road most rewarded. Life is unpredictable, the world is big and complicated, and this is just the beginning. In college, in graduate school, at work, and in personal life, the 17-year-old will have to continue to decide which actions to take as he/she grows up, learns to find his/her place, and seeks to build a meaningful and rewarding life, in a world fraught with dangers.

How to learn to make good decisions? This is the topic of Reinforcement Learning (RL).

> *Reinforcement learning solves sequential decision-making problems.*

7.1 Reinforcement Learning

The underlying idea is to make an agent navigate an environment in a way which maximizes overall reward. The agent's state changes as it takes actions, and the goal is to learn how to navigate the environment by taking the best actions. The goal could be very specific, such as walking, running, and playing, with a given set of possible actions we can take in order to accomplish the task. See Fig. 7.1.

The knowledge about which actions work best for which state is called *policy*. Policy can be probabilistic or deterministic, and it has to be learned gradually during exploration of the environment. Based on the formulation of the learning problem, a slew of algorithms have been discovered to solve the problem.

A model of the environment may be known, unknown, or partially known, and it may be finite or infinite. An agent may be able to observe the state of the environment fully or partially. The world may be discrete or continuous. The agent may choose to apply the learned policy, while it explores the world. Alternatively, it may choose to move around and take actions using a different policy, even a random one, while it learns. Therefore, in research papers, we talk about model-based or model-free learning, on-policy or off-policy learning, discrete or continuous states, discrete or continuous actions, full or partial observations, behavioral or learned policy, and deterministic or stochastic policy.

Certain terms are used in RL which can be easily related to human life. *Reward* is the most important one, and it is the incentive behind learning. Reward is formalized as something one gets immediately on taking an action in a given state. The trajectory, rollout, or episode of learning involves going from state to state, till a terminating state is reached. The rewards accumulate over time, and we may adopt a short-term horizon, giving importance to immediate rewards while discounting rewards in the future. Some may exercise patience and prefer a long-term horizon. If one loses out on reward because one is unable to follow the optimal policy, it is called regret. Finally, given any state, the advantage of an action over other actions indicates the relative value of that action over other actions on average. It is the difference between the reward one gets by taking that particular action over taking alternative actions according to a policy, assuming one continues to follow the same policy.

We apply what we learn and this is called exploitation of our knowledge. Suppose you have learned that it pays off to be a specialist and to stick to one's field and you live out life with this policy. But there is more to life and the world, and getting out of one's comfort zone is exploration in RL. It is implemented as an ϵ-greedy algorithm, in which the agent chooses to explore with ϵ probability. It allows the agent to depart from the policy occasionally and explore new options. A fine balance between exploration and exploitation is needed for the maximization of rewards.

Fig. 7.1 Humans use reinforcement learning to learn goal-driven multistep tasks. Robots are trained for the same tasks in a similar way

RL uses concepts intuitively very similar to how people make decisions in their life trajectories and in handling an uncertain world and future. A key concept is that of policy, which guides the agent in choosing the best action in a given state.

7.2 Learning an Optimal Policy

We start with the simplest scenario. Call it a pre-RL scenario because everything about the world is given in the form of a directed graph of states connected by actions called a Markov Decision Process (MDP), which has been studied since the 1950s. In each state, a set of actions is available. Given a state and an action, the agent lands stochastically in other states as specified by transition probabilities. A reward is obtained when this transition occurs. The goal is to find a policy, which tells the agent which action to choose in every state. See Fig. 7.2.

A key concept employed is the *value* of a state, which is the expected accumulated reward to be obtained by landing in that state and continuing the journey onward. Therefore, the value of a state in which you are married to the right person or in which you are studying a subject which you feel enthusiastic about is high, because the expected rewards are larger than being in a state which is not desirable.

An MDP has to observe self-consistency between the optimal values of different states because states are interconnected. This self-consistency is given by the *Bellman equation*:

$$V(s) = \max_{a}(r_{s,a} + \gamma V(s')),$$

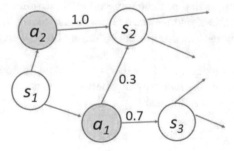

Fig. 7.2 A portion of a Markov decision process. At each state, there are choices of actions to take. The transition probabilities determine the next state when an action is taken. The goal is to find the best action in a given state, which maximizes the expected cumulative reward

where s is the old state, s' is the new state, $r_{s,a}$ is the reward obtained by taking action a in state s, and γ is the discount factor. It defines the value of a state as the maximum cumulative reward when starting from that state in terms of values of the next states. Therefore, the value of a state in which you are studying the right subject in college and the value of a future adjacent state in which you are working in that area have to be consistent.

A deterministic policy π is a function which maps states to actions,

$$\pi : S \rightarrow A,$$

where S is the set of states and A is the set of actions. A trajectory is a rollout of (state, action) pairs according to the policy,

$$s_0, \quad a_0 = \pi(s_0), \quad s_1, \quad a_1 = \pi(s_1), \quad s_2, \quad \ldots,$$

where the actions are selected by the policy and the next states are determined by the state transition probabilities $T(s_{t+1}|s_t, a_t)$, in a stochastic manner. The optimal policy π^* maximizes the expected cumulative reward starting from the first state s_0.

When we have complete information of the world, the Bellman equation can be solved using convergent iterative methods, which update the values and the policy. The state values are updated from the present estimate of the policy and vice versa. It doesn't matter whether these updates occur simultaneously over all states or in some order, state by state. The algorithms converge to the optimal policy π^* and the optimal state values $V^*(s)$, and MDP becomes a Markov chain. The intuition between the iterative updates of the state values and the policy is to make them self-consistent and satisfy the Bellman equation.

The problem is more challenging if the complete picture of the world in the form of an MDP is absent. In simulation, it is like playing an open-world video game, which has not been explored fully. In the future, it will be like exploring a new planet, which we are unfamiliar with. We don't know everything. We have trajectories, which unfold in time depending upon our actions; we don't know about the transition probabilities; and we learn about rewards with time. In such cases of unknown worlds waiting to be discovered, solution of model-free MDPs is provided by RL.

One of the earliest model-free RL methods is based on Temporal Difference (TD) error, which is defined as the *inconsistency* between the actual state value and what is expected according to the Bellman equation,

$$\text{TD Error} = r + \gamma V(s') - V(s).$$

If the TD error were to be zero, the Bellman equation would be perfectly satisfied. A TD-based method minimizes the TD error. Such a method was famously used in a program that learned to play the game of backgammon at par with expert human players. Note that TD methods combine Monte Carlo sampling with dynamic programming.

1. The Monte Carlo algorithm samples trajectories stochastically according to the policy and the transition probabilities.
2. Dynamic programming reduces the TD error by reusing the previously computed state values and policy.

The intuition is to iron out the mutual inconsistencies between the state values while one explores the environment.

Let us refine the idea further. Define Q-values for state-action pairs. The Q-value of an (s, a) pair is the expected cumulative reward on taking action a in state s, if the agent continues to follow the same policy. Consider the following algorithm.

1. Use the Monte Carlo algorithm to sample trajectories according to an ϵ-greedy behavioral policy, which strikes a trade-off between exploration and exploitation. The exploration probability is ϵ, and the exploitation probability is $1 - \epsilon$. In exploitation, actions are selected based on the current policy, that is, the action that maximizes $Q(s, a)$ is chosen in a greedy step. In exploration, actions are selected randomly.
2. Compute the TD error,

$$\mathcal{E} = r + \gamma \max_{a'} Q(s', a') - Q(s, a).$$

3. Reduce the TD error according to a learning rate,

$$Q(s, a) = Q(s, a) + \alpha \mathcal{E},$$

where α is the learning rate. Note the similarity with a gradient step in SGD. Verify that the above reduces the error.

This is Q-learning, which will converge to

$$Q(s, a) = \mathbb{E}_{s'} \left[r + \gamma \max_{a'} Q(s', a') \right].$$

Note that Q-learning is off-policy because it uses an ϵ-greedy behavioral policy to move from state to state, which is different from the policy being learned. It uses exploration to ensure that there is a good coverage of the state-action space. If $\epsilon = 1$, then it decouples itself completely from the policy being learned, and it explores all the time. However, exploring all the time will lead to significant slowdown when the state space is very large. This is contrasted with an alternative approach as follows.

1. Use the Monte Carlo algorithm to sample trajectories according to the current policy and the transition probabilities.
2. Compute the TD error,

$$\mathcal{E} = r + Q(s', a') - Q(s, a),$$

   where a' is the next action in the trajectory as given by the current policy.
3. Reduce the TD error according to a learning rate,

$$Q(s, a) = Q(s, a) + \alpha \mathcal{E},$$

   where α is the learning rate.

The above algorithm is on-policy and is acronymed as Sarsa, standing for the quintuple – state, action, reward, state, action.

*** Exercise 19   Sarsa/Q-learning

Suppose you are trying to navigate a world full of unknown dangers such as abysses, fire, quagmires, big spiders etc., in search of treasures, and you use two algorithms A and B to find the optimal policy. Algorithm A returns a near-optimal policy, which seems to keep you safer by keeping you away from the dangers. Given that one algorithm is Sarsa and the other is Q-learning, specify which of A and B is Q-learning?

Once the Q-values have converged, then one can find the best policy according to them. The policy will be deterministic, and one simply chooses the action with the highest expected reward. Note that Q-learning is really designed to learn deterministic policy. In this chapter, we will see soon how one directly optimizes a stochastic policy on a large state space.

The Bellman equation for optimality in a state-action space is the basis for RL algorithms, which seek to satisfy the self-consistency implied by the equation. In Markov decision processes, where everything is known in advance, the Bellman equation can be satisfied exactly. In a partially known world, the Monte Carlo algorithm is used to explore the world, and the Q-values are updated in order to satisfy the Bellman equation.

7.3 Deep Q-Learning

TD-based RL algorithms, such as Q-learning and Sarsa, need to keep a table called a Q-table; see Fig. 7.3. The size of the table is $N \times M$ for M states and N actions, and it is impractical for real-world applications. Imagine counting the number of states in the game of Go and keeping a Q-table covering all the states.

The idea that makes Q-learning work for real-world applications with large state spaces is to replace the table with a function. The function is implemented by a DNN, which can compute the Q-value for any cell of the table. The DNN is called a Deep Q-Network (DQN). We say that the Q-function has been parameterized by the DQN. Deep Q-learning was popularized by DeepMind in building an AI, which learns to play Atari video games from scratch given the raw pixels of game screens. The state space consists of the images on the computer screen in the games.

A DQN is trained to minimize the TD error using the standard backpropagation algorithm. The loss function is the expected squared TD-error,

$$\mathbb{E}[(r + \gamma \max_{a'} Q(s', a') - Q(s, a))^2],$$

where the expectation is over the transitions s, a, r, s' collected while executing the behavioral policy. In practice, instead of taking the expectation by exploring the state space, which has the side effect of creating correlations between consecutive samples, experience replay is used in which the agent's trajectories are saved in a replay memory from which random samples are taken. The loss and its gradient are computed by using a minibatch of transitions sampled from the replay memory. This breaks the correlation and improves convergence of the training process.

Fig. 7.3 Q-table. Each cell of the table contains the current estimate of the Q-value. The method is impractical for real-world applications with large state spaces. Q-learning updates the Q-values in the direction of minimizing the squared TD error

Q(s,a)	s_1	s_2	s_3
UP		0.3	
DOWN	0.4		
LEFT			-0.5
RIGHT		1.5	

*** **Exercise 20   DQN**

See [77]. State how double Q-learning works and its effect on the performance of the algorithm.

An improvement of the DQN method, which is frequently applied in practice, can be mentioned at this point. In this variation, in the computation of the TD error,

$$r + \gamma \max_{a'} Q(s', a') - Q(s, a),$$

Q-values are computed using two DNNs, the policy network P and the target network T,

$$r + \gamma \max_{a'} T(s', a') - P(s, a).$$

T periodically updates its parameters from P and keeps them frozen in between for stability and better convergence of the algorithm. Otherwise, if the target network is the same as the policy network, then there is a feedback loop, which is often unstable in practice, since the current output of the policy network affects the backpropagation in the next step. The feedback happens because the same network sets its target Q-values and then strives to reach them. It is often described as a dog chasing its own tail.

Deep Q-learning replaces the quality table in Q-learning by a DNN. The deep Q-network is trained by minimizing the L^2 norm of the temporal difference errors on transitions sampled from an experience replay memory.

7.4   Policy Gradient

Despite the fact that DQN gained fame in 2013 by making AI play video games, another method known as policy gradient has turned out to be more practical, versatile, and powerful,l and it has led to the remarkable success of AlphaGo in its match against the genius Lee Sedol. This method can learn stochastic policy directly, where it outputs a discrete probability distribution over a discrete action space which can be high-dimensional. It can also work with continuous action space. For a continuous action, it outputs the parameters of a continuous probability distribution from which the action can be sampled. In contrast, DQN works with discrete low-dimensional action space for which a deterministic policy is learned.

As for any AI model, for both DQN and policy gradient, the hypothesis is that, for different actions, different regions of the state space can be carved in terms of convex polytopes by the network. The success of AlphaGo is ultimately based on enormous carving capacity of AI models.

7.4.1 Intuition

The goal is to learn a stochastic policy directly in a large state space such as the game of Go or a video game. A stochastic policy is preferred over a deterministic policy in competitive games or in large state spaces where there is uncertainty. A deterministic policy is not the best choice, because we never know everything. We use a DNN which maps the states to the stochastic policy; it is called a Policy Network (PN).

Let there be N actions in the problem space. If you are playing a game of Snake, in which the snake gets longer over time, $N = 4$ corresponding to the four movement directions of the mouse cursor on the computer screen. For Snake, the PN model takes a configuration of the game screen as the input and outputs a four-dimensional stochastic policy, that is, probabilities of the four actions Up, Down, Left, and Right:

$$[\pi_\theta(\text{Up}|X), \ \pi_\theta(\text{Down}|X), \ \pi_\theta(\text{Left}|X), \ \pi_\theta(\text{Right}|X)] = \pi_\theta(X),$$

where X is the game screen as an image, π_θ is the function implemented by the PN model, and θ are the trainable parameters of the DNN. Let's formulate the problem we are trying to solve.

1. We run the Monte Carlo algorithm according to the current policy to roll out trajectories and get cumulative rewards. In the case of a game such as Go, the reward is given at the end of a trajectory in terms of Win, Lose, or Tie. At each time step, an action is sampled according to the stochastic policy modeled by the PN model. This continues till a terminating state is reached and a reward is received, which may be positive or negative. The probability of a trajectory is the product of the probabilities of the state transitions and action selections made along the trajectory. See Fig. 7.4 for a visual illustration of a trajectory.
2. The expected reward over many trajectories will be a weighted sum of rewards, where the weights are the trajectory probabilities. The problem can be formulated as the following: find a PN model which outputs such action selection probabilities that maximize the expected reward. For a positive reward, the PN model needs to be trained to increase the probabilities of the chosen actions taken along the trajectory. For a negative reward, the goal is to decrease the probabilities. Therefore, the PN model is being provided a ground truth for backpropagation. Figure 7.4 shows the objective function which the PN model needs to maximize at every time step.

To give an intuition behind the solution before working out the mathematical details, assume that action a is selected at a time step, and recall that for classification the

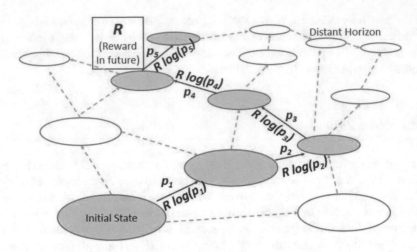

Fig. 7.4 The policy gradient method. A Monte Carlo trajectory is shown by the red arrows in a state space, which samples actions given by the stochastic policy. The probabilities of the chosen actions are shown on the edges. It terminates in a final state, where the reward R is obtained. Then, the policy network is trained to maximize the objective function shown on the edges, that is, $R\nabla \log p_i$ is the policy gradient. If $R > 0$, then the probabilities of the chosen actions are increased. If $R < 0$, then they get decreased

loss is $-\log p_a$ if the "ground-truth classification label" is the action a. What if we were to modulate it with the final reward of Win, Lose, or Tie? Suppose the rewards are as follows:

$$R(\text{winning trajectory}) = +1$$

$$R(\text{losing trajectory}) = -1$$

$$R(\text{tied trajectory}) = 0.$$

Therefore, the proposal is to scale or modulate $-\log p_a$ with R,

$$-R \log p_a,$$

which has to be minimized, or equivalently, to create an objective function which is maximized,

$$R \log p_a.$$

For a winning trajectory, the backpropagation algorithm will adjust the learnable parameters of the PN model toward the direction of making the action more likely, and for a losing trajectory, less likely. For a tied trajectory, no adjustment is made.

This has to be done over a large number of trajectories, which should ensure that on average the reward is maximized. The above proposal of a reward-modulated objective function is based on intuition. We will now show that this proposal is indeed mathematically sound and that it will indeed maximize the expected reward when applied on many trajectories.

7.4.2 Mathematical Analysis

This section is written for readers who wish to dive deeper into the mathematics of the policy gradient method.

Suppose a Monte Carlo trajectory τ_1 starting at an initial state s_0 is the sequence

$$s_0 \xrightarrow{\pi_\theta} a_0 \xrightarrow{T} s_1 \xrightarrow{\pi_\theta} a_1 \xrightarrow{T} s_2 \rightarrow \ldots \rightarrow a_{N-1} \xrightarrow{T} s_N \xrightarrow{\text{Reward}} R_1,$$

where the reward R_1 is obtained in a terminating state s_N. The reward can be cumulative along the trajectory and in that case one just uses the cumulative reward. The probability of τ_1 is the product of the action selection probabilities according to the policy π_θ and of the state transition probabilities,

$$P(\tau_1) = \prod_k \pi_\theta(a_k|s_k)T(s_{k+1}|s_k, a_k).$$

The Monte Carlo algorithm is executed to yield M simulated trajectories in total,

$$\tau_1, \tau_2, \ldots, \tau_M,$$

which lead to rewards

$$R_1, R_2, \ldots, R_M,$$

where M is a large number. The objective function to be maximized is the expected reward

$$F(\theta) = \mathbb{E}_\tau[R] = \sum_{i=1}^{M} P(\tau_i)R_i.$$

To train the PN model, the objective function is differentiated with respect to θ. The first step in the chain rule is to differentiate the objective function with respect to the $P(\tau_i)$ term, which depends on the outputs $\pi_\theta(a_i|s_i)$ of the PN model at each time step, which in turn depend on the internal learnable parameters θ of the PN model. The gradient of $F(\theta)$ with respect to θ is called the policy gradient,

$$\nabla_\theta F(\theta) = \nabla_\theta \mathbb{E}_\tau[R] = \sum_{i=1}^{M} \nabla_\theta P(\tau_i)R_i.$$

From high-school calculus, we know that

$$\frac{df(x)}{dx} = f(x)\frac{d \log f(x)}{dx}.$$

Therefore, we can rewrite the policy gradient as

$$\nabla_\theta F(\theta) = \sum_{i=1}^{M} P(\tau_i)\nabla_\theta \log P(\tau_i)R_i,$$

which reduces to the expectation

$$\nabla_\theta F(\theta) = \mathbb{E}_\tau[\nabla_\theta \log P(\tau_i)R_i].$$

Recall that $P(\tau_i)$ is a product of probabilities and therefore its logarithm is a sum of probabilities, which can be differentiated. Consider τ_1 again,

$$\log P(\tau_1) = \sum_k (\log T(s_{k+1}|s_k, a_k) + \log \pi_\theta(a_k|s_k)).$$

In taking the gradient, the state transition probabilities are independent of θ, and therefore they can be discarded, leading to the following expression,

$$\nabla_\theta \log P(\tau_1) = \sum_k \nabla_\theta \log \pi_\theta(a_k|s_k)$$

and therefore, the policy gradient for the trajectory τ_1 is obtained by modulating each term with the reward R_1,

$$G(\tau_1) = \sum_k \nabla_\theta \log \pi_\theta(a_k|s_k)R_1.$$

Taking the expectation over M trajectories means running a Monte Carlo algorithm to generate M trajectories and then taking the average of the above,

$$\nabla_\theta F(\theta) = \frac{1}{M}\sum_{i=1}^{M} G(\tau_i).$$

This is the final expression. Therefore, the policy gradient method is the following algorithm.

1. Run the Monte Carlo algorithm to generate M trajectories.
2. For each trajectory and for each time step in the trajectory, set the objective function to the *reward-modulated logarithm of the probability of the sampled action* and perform backpropagation to *maximize* it.

Consider again the example of sampling from the policy for the Snake game,

$$[\pi_\theta(\text{Up}|X), \ \pi_\theta(\text{Down}|X), \ \pi_\theta(\text{Left}|X), \ \pi_\theta(\text{Right}|X)] = \pi_\theta(X).$$

If "Up" was sampled in the Monte Carlo algorithm, then having "Up" as the ground-truth label for that time step, $R \log \pi_\theta(\text{Up}|X)$ is the objective function to be maximized or, equivalently, $-R \log \pi_\theta(\text{Up}|X)$ is the loss function to be minimized. This is exactly like minimization of cross-entropy loss for a positive sample except that it is modulated by R. See Fig. 7.4, where the policy gradient is

$$R(\nabla \log p_1 + \nabla \log p_2 + \ldots + \nabla \log p_5).$$

An important thing to keep in mind is that the above loss function is not really a "loss function" in the conventional sense. This function may improve with backpropagation, without improving the final objective function, which is the expected reward. It is just a function defined on the current trajectory data obtained from the current parameters whose gradient is helpful on the current data in a stochastic sense.

The mathematics of the policy gradient method can be extended further to make the training stable, because in its original form, the policy gradient method is known to suffer from high variance. Several alternatives weighting factors besides $R(\tau)$ exist. For example, the reward function $R(\tau)$ can be replaced by the quality function $Q(s, a)$. In turn, $Q(s, a)$ can be approximated by evaluating whether taking an action a is advantageous. An approximate $Q(s, a)$ can be predicted by a separate DNN called a critic. The PN model is then referred to as an actor, and the whole scheme is called an actor-critic method. The actor doesn't have to wait for the final outcome $R(\tau)$ of the trajectory, because the critic immediately provides the feedback.

> *The policy gradient method uses the SGD algorithm to train an AI model, which computes policy. The AI model is known as the policy network. The loss function includes a multiplicative factor of the cumulative rewards of the Monte Carlo trajectories, so that the parameters of the policy network are nudged in the direction of maximizing the expected reward.*

7.5 Let's Play and Explore

Deep Reinforcement Learning (DRL) has helped AI win at games. Behind these victories, it takes a significant effort to accomplish these feats and a lot more goes into the final solution than just using a DNN for a policy network. For noteworthy

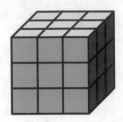

Fig. 7.5 Rubik's cube and
Go can be solved by
employing deep
reinforcement learning. A
number of innovative
techniques have to be
carefully put together to make
AI excel in game playing

work, see [196–198]. The final AI model is a result of a combination of several
techniques painstakingly put together as components of an overall solution.

A useful technique is Monte Carlo Tree Search (MCTS), a generalization of
Monte Carlo methods for single-player games to competitive games. In a two-player
competitive game, a tree of trajectories is unrolled using the MCTS algorithm. This
is done by making the AI play against itself in *self-play*. In such Monte Carlo
rollouts, the underlying idea is that though we can't tell conclusively whether a
particular state-action pair is beneficial or not when playing a few games, we can
infer so by simulating a large number of games. If it is beneficial on average over
many games, then it is beneficial. The analogy is that if a particular action, such as
going to college, has worked well with many people, then it is likely that you will
also benefit from it.

A combination of several techniques, which are fine-tuned with diligence, led to
the success of AlphaGo. In fact, AlphaGo is bootstrapped from a large database of
games played by expert Go players in training its policy network using supervised
learning, which also involves game-specific features. The value network, which
predicts whether a state is good for winning or not, is trained using RL. The
policy network prunes the edges of the self-play search tree, and the value network
evaluates the leaf nodes at a particular truncating depth. Both networks cut down
the enormous 3^{361} (a Go board is 19×19) search space of all board configurations,
approximately 2×10^{170} of which are legal.

In AlphaZero, the dependence of the policy network on the database of games
is eliminated through extensive use of self-play and iteratively improving the two
networks starting with random initialization. In each iteration, a vast number of
game trajectories are rolled out using self-play guided by the current policy and
value networks. This allows the algorithm to generalize to chess and shogi.

Self-play works well but it is valuable to learn from experts. Why learn
everything from scratch? We can complement RL techniques by *imitation learning*,
in which an AI model gets supervised data by observing an expert. This is also
known as apprenticeship learning. Therefore, in order to learn race car driving, one
can observe an expert driver and obtain a training dataset of (s, a) pairs. One doesn't
have to wait for the reward to learn policy.

Also see [3] to learn how DRL can be used to solve Rubik's cube; see Fig. 7.5. To
make AI learn efficiently where rewards are sparse and rollouts may not terminate,
a technique called autodidactic inference is used. A combined value and policy
network $(v, \pi) = F(s)$ is trained. From the terminating state, other states are
generated in a backward direction,

$$s_t \leftarrow s_{t-1} \leftarrow s_{t-2} \leftarrow \ldots.$$

For each generated state s, its children obtained by taking different actions (turning moves) are evaluated using the current network F, and the maximum value among its children leads to the new value and policy training targets for the state s. Once the network is trained, then it is used inside an MCTS-based rollout algorithm to solve the puzzle starting with an initial state s_0.

Let us look beyond games for applications of RL. Wherever there is a sequential decision-making process, RL can be applied. Robotics, self-driving cars, self-flying aerial vehicles, business strategy, inventory management, industrial design, logistics and operations, wellness, preventive medicine, medical treatment planning, trading strategy, supply chain management, and education can potentially benefit. Moreover, neural architecture search, NLU tasks, and recommender systems can be formulated in terms of RL. Anywhere you need to make *state-based decisions*, consider RL. Anywhere you need to search efficiently in a dynamic state-based domain, think of RL. And, the problem need not be a practical one. Consider RL formulation even for theoretical research such as for NP-hard or combinatorial optimization problems.

To provide an intuition about how RL can be used in the real world, consider the case of treatment planning in medicine, specifically in radiotherapy. A treatment planner uses certain dosage hyperparameters and the anatomy of a patient to produce a treatment plan using a treatment planning software. The treatment plan is a 3-D dose distribution, which is often represented compactly with 1-D dose volume histograms, over the target organ and proximal organs known as organs at risk. One should provide the prescribed radiation dose to the target organ and spare the organs at risk. Different organs at risk will have different sensitivities to the radiation. Based on the consultations with the oncologist and the evaluation of the treatment plan, the planner will iterate several times to tweak the hyperparameters in order to produce a better quality treatment plan. This is an extremely labor-intensive process. We can formulate this as an RL problem. The treatment plan is analogous to a game configuration. The actions are tweaks of the treatment plan parameters. The reward function is based on certain objective criteria, which evaluate how much dosage the target organ and the organs at risk receive. A policy network takes a treatment plan as the input and outputs a distribution over the updates of the plan parameters. The updated plan parameters are then used to create a new treatment plan, and the whole process is repeated. The final treatment plan will be of higher quality than the initial proposed plan. This is just for one treatment session. For treatment which occurs over many sessions spread over time, this approach can be generalized where one moves from one session to the next and adapts to any changes in the anatomy and other clinically relevant environment. For such an important application of RL, see [194, 215, 222].

In a futuristic world, we will gift robot friends to our children. Such a robot will arrive in a box delivered to the doorstep by a robot seated in a self-driving car. On unpacking the box, the robot will stumble out of the box, gradually learning to walk and run, just like a human baby using RL. With time, it will read books to the children. It will learn how to tell stories, a sequential process of steps of narration intermixed with animated gestures and facial emotions, with the overall "reward" of making our children experience the magic of stories.

> *Playing Go and solving Rubik's cube show the effectiveness of deep reinforcement learning in solving sequential decision-making problems in constrained domains. In the future, we will witness diverse applications of DRL in more general domains.*

7.6   Chapter Summary

Reinforcement learning is used to make machines learn in a state-based, interactive environment. A central concept is that of policy, which guides the machine in choosing actions which maximize the expected cumulative reward. If everything about the environment is known in advance, then the problem is formulated as a Markov decision process, which can be solved exactly by applying the Bellman equation iteratively to reduce the temporal difference error. The Bellman equation expresses self-consistency constraints which a Markov decision process has to respect. The temporal difference measures departure from this desired self-consistency. When there is partial knowledge of the system, then reinforcement learning is used to learn policy. Q-learning is a reinforcement learning algorithm, which keeps a Q-table for (state, action) pairs. The table is updated to reduce the temporal difference error using learning steps controlled by a learning-rate hyperparameter. When the table is implemented by an AI model, then it is called deep Q-learning. The deep Q-network is trained by minimizing the temporal difference error. In practice, the policy gradient method is used, in which the policy is implemented by an AI model known as the policy network. The Monte Carlo rollouts are performed, and the reward values are propagated back to train the policy network to maximize the objective function of the expected cumulative reward over the rollouts. The gradient of the objective function of the expected reward is the same as the gradient of the sum of the logarithms of the probabilities of the selected actions, and these logarithm terms are modulated by the rewards obtained in the rollouts. This gradient is then backpropagated to the learnable parameters of the policy network. Reinforcement learning has led to the success of AI in game playing. At the same time, it has potential applications in many domains.

Problem Set 7

1. In general, what are Monte Carlo algorithms?
2. What is exploration-exploitation trade-off?

3. What will be a good strategy to change ϵ in the ϵ-greedy policy for exploration-exploitation trade-off?
4. What does a large value of the discount factor γ indicate?
5. Suppose $\gamma = 0.9$ and the immediate reward is 10 followed by an infinite sequence of rewards of 5. What is the total discounted award?
6. Suppose you are training an RL model with the discount factor $\gamma = 1$. You notice that the agent takes a very long time to reach the goal where there is a big reward. How will you train the agent to reach the goal faster?
7. In an MDP, is the optimal state-value function unique?
8. If one adds a constant to the reward function, would the optimal policy change?
9. What is the biggest limitation of Q-learning?
10. Why does the policy gradient method work so well for stochastic policies?

Chapter 8
The Classical World

> A classic is a book that has never
> finished saying what it has to say.
>
> ———————————————
>
> Italo Calvino

The topic of this book is AI, which is rooted in classical ML. In order to understand AI better, it is important to revisit the classical techniques. This chapter will take you on a brief journey through the ideas and methods, which laid the foundations of ML. The fundamental difference from AI is that one is working in a carefully designed feature space.

The feature-engineering process involves meticulous and intensive work to compute features, which have to be researched, visualized, selected, and implemented. A classical ML algorithm will approximate the desired function or separate out the given classes in the feature space.

Classical ML is based on computer science, statistics, and mathematics. It offers many techniques, which are used in AI. AI has complemented these time-tested techniques with new ones, such as ReLU activation, max pooling layers, variations of the SGD algorithm, and novel architectures. Coupled with huge advances in computing power and the availability of large datasets, the classical ML has evolved into AI.

© The Author(s), under exclusive license to Springer Nature Switzerland AG 2021   159
S. Dube, *An Intuitive Exploration of Artificial Intelligence*,
https://doi.org/10.1007/978-3-030-68624-6_8

8.1 Maximum Likelihood Estimation

For probability, we look at the future based on what is happening now and ask how probable a future outcome is. For likelihood, we are looking at the past based on present observations, and we ask how likely a past causal event was. Given the observed data D, which is an "effect" of something that happened in the past, we can estimate the probability of this effect given a particular "cause."

The Maximum Likelihood Estimation (MLE) principle states that we should estimate parameters θ such that the probability of observing the data is maximized:

$$\hat{\theta} = \arg\max_{\theta} \Pr(D|\theta).$$

The conditional probability term is written as the likelihood function $L(\theta|D)$, where θ is variable and D is fixed.

Consider ten observations of random tosses of a coin,

$$H,H,T,H,T,H,H,H,T,H,$$

where we need to estimate $\theta = \Pr(H)$. We can estimate it to be $\hat{\theta} = 0.7$ because we see seven heads and three tails. This is the MLE because it maximizes

$$L(\theta|D) = \Pr(D|\theta) = \theta^7(1-\theta)^3,$$

where we can quickly differentiate it with respect to θ for confirmation.

The logistic regression method is an example of techniques, which are based on the MLE principle. Moreover, the regression models apply the MLE principle in the presence of noise. We will see that the type of noise determines the type of L^p loss needed for maximum likelihood maximization. AI models for classification and regression apply the MLE principle just like the classical ML techniques. Therefore, this estimation technique provides a foundation for major techniques in AI and ML. The MLE principle is a special case of Bayesian estimation, which we will discuss in the next section, when the prior probability distribution of the parameters to be estimated is noninformative.

*** **Exercise 21**   **Estimation**

Consider a constant stochastic process, which emits a value x, such that

$$x = \theta + \text{Noise}.$$

State the MLE of θ given observations $D = \{x_1, \ldots, x_N\}$ for the following two cases:

1. Noise is Gaussian.
2. Noise is Laplacian.

Specify the constant functions fitted to D, which satisfy the following:

1. Minimum L^2 error or mean squared error
2. Minimum L^1 error or mean absolute deviation

8.2   Uncertainty in Estimation

Intuitively speaking, the MLE principle makes sense. At the same time, chance plays a role, and the estimate may be off from the true value. Statistics comes to our rescue in quantifying this uncertainty due to chance.

In the coin toss example in the previous section, suppose the true value of $\Pr(H)$ is 0.5, and we saw the observed sequence purely by chance. If we were to repeat the experiment of N coin tosses many times, the mean of the estimates would converge to the true value of 0.5. These estimates form a probability distribution with the mean at the true value and a standard deviation of $O(1/\sqrt{N})$. This is the foundation of hypothesis testing and confidence intervals in classical statistics. When we talk about a 95% confidence interval, it means that if we were to repeat the experiment, the true value would be in the confidence interval 95% of the time. The null hypothesis asserts a value for an unknown parameter, and an experiment is done to test it. The p-value is the probability of observing the outcome or a more extreme outcome away from the asserted value, provided the null hypothesis is the cause. If the p-value is below a critical threshold such as 5%, we reject the hypothesized cause and accept the alternative one. Note that the underlying parameter is assumed to be a constant in the hypothesis testing.

In Bayesian statistics, the set of "causes" is endowed with a probability distribution known as the prior distribution, and, given an "effect," the conditional probability of a particular cause known as the posterior probability,

$$\Pr(\theta|D) = \frac{\Pr(D|\theta)\,\Pr(\theta)}{\Pr(D)},$$

is computed, which is known as the Bayes' theorem and can be written as

$$\text{Posterior Probability} = \frac{\text{Likelihood} \times \text{Prior}}{\text{Evidence Probability}}.$$

To maximize the posterior probability, we maximize the numerator since the denominator is constant. If the prior is uniform, then it reduces to the MLE principle. See Sect. 15.1.2 and Exercise 33, where we use Bayesian inference. The posterior probability of a cause is our current belief in the likelihood of the cause. The power of Bayesian inference lies in an iterative update of the beliefs starting with the prior. As new evidence shows up, the belief is iteratively updated. This is called Bayesian inference. Note that given a cause A and an effect E, the Bayes inference rule can be rewritten as

$$\Pr(A|E) = \frac{\Pr(E|A)\,\Pr(A)}{\Pr(E|A)\,\Pr(A) + \Pr(E|\overline{A})\,\Pr(\overline{A})}.$$

> *Maximum likelihood estimation and Bayesian inference provide the basis for a number of classical ML tools.*

8.3 Linear Models

8.3.1 Linear Regression

See Sect. 1.4. We generalize Exercise 21 to unravel linear regression. Consider a linear stochastic process, which emits a value y, such that

$$y = \theta_0 + \sum_{i=1}^{k} \theta_i x_i + \epsilon,$$

where ϵ is noise. Then, the MLE estimate of the θ_i's given a training set of k-dimensional vectors is as follows:

1. If ϵ is Gaussian, then the solution is given by Ordinary Least Squares (OLS).
2. If ϵ is Laplacian, then the solution is given by Least Absolute Deviation (LAD).

Alternatively, in order to fit a linear function to the training data under different loss functions, we have the following results:

1. The OLS method will minimize L^2 error. Equivalently, it minimizes the mean squared error or sum of squares error.
2. The LAD method will minimize L^1 error. Equivalently, it minimizes the mean absolute deviation or sum of absolute deviations.

The OLS method has direct analytical solutions based on linear algebra. The fitted line passes through the mean of the data. Consider the line-fitting case where $k = 1$ for the LAD method. Like in the case of the constant stochastic process, which latches to the median value, see Exercise 21 and its solution, and is therefore robust to outliers, the fitted line latches to pairs of data points in the middle, which may be nonunique, resulting in multiple solutions, and robust to outliers.

8.3.2 The Geometry of Linear Regression

Let's develop geometrical intuition behind the OLS and LAD methods for the 2-D case. The loss function is a surface over the 2-D plane. Let $y = \theta_0 + \theta_1 x + \epsilon$. Consider a loss function over training data. For any constant C, define the corresponding contour of the loss surface to be the set of those parameter values for which the loss is C. Then, for the L^2 loss function, the contours of the loss surface are conics, and, for the L^1 function, they are piecewise linear.

8.3.3 Regularization

Regularization is optimization with constraints and is solvable by the Lagrange multiplier method. Regularization is a prior belief, which we want to incorporate in the process. Often we believe that the model parameters should have small magnitude because simpler models are to be preferred. The justification of this belief is that complex models are likely to overfit, and therefore, regularization is a variance-reducing method. The loss function for regularized linear regression is given by

Regularized Loss Function $= (L^p \text{ error}) + \lambda * (\text{Size of Coefficients})$,

where λ is the Lagrange multiplier. There are several choices for "size of coefficients," and it could be the L^1 norm (known as the lasso penalty) or the L^2 norm (known as the ridge penalty) of the parameter vector. The lasso penalty defines concentric diagonally oriented squares as the constraints, and the ridge penalty defines concentric circles. The value of λ defines the largest valid region for the parameters, that is, the largest square or largest circle, respectively, within which the L^p error has to be minimized. See Fig. 8.1 to develop intuition. The ridge and lasso penalties can be linearly combined; this is called elastic net regularization.

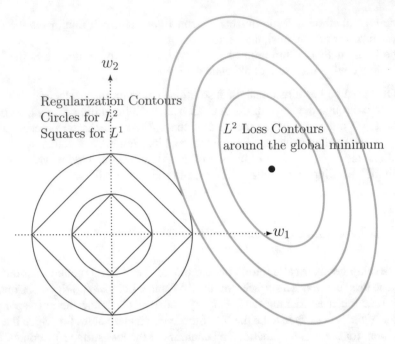

Fig. 8.1 We want to optimize a convex function $f(x_1, x_2)$ defined over a 2-D parameter space. Its contours are shown in green and enclose the global minimum. For the L^1 loss, contours will be piecewise linear. Adding a regularization term $g(x)$ means we are minimizing $f(x) + \lambda g(x)$, where λ is the Lagrange multiplier. If $g(x)$ is the L^2 norm of the parameters, then the contour plots of $g(x)$ are blue circles, for varying values of λ. The optimum will be where a green contour touches the blue circle for the given λ. For L^1 norm regularization, a green contour has to touch a red square

> *Linear models with regularization can be understood geometrically, and they constitute the most fundamental classical ML techniques.*

8.3.4   Logistic Regression

Logistic regression is a linear model for classification. A logistic regression module is obtained by attaching the sigmoid activation function to a linear regression module, which squishes the output to [0, 1] so that it is a probability value, and by using the cross-entropy loss for binary classification. The logistic regression method applies the MLE principle to estimate parameters. See Sect. 1.4 for intuition behind the cross-entropy loss.

8.4   Classical ML

We briefly survey a few well-known classical methods at a high level. For certain applications, these methods will work fine.

8.4.1   k-Nearest Neighbors

A nonparametric approach is the k-nearest neighbors (k-NN) method. Given an unseen example x, its k-nearest neighbors are found in the feature space, and their labels and attributes are used to predict those of x. As k increases, the variance reduces. In high dimensions, we have to resort to approximate algorithms to find the nearest neighbors. It can be shown that there is no quick way of doing exact nearest neighbor search as the number of features increases. In low-dimensional cases, data structures such as k-d trees can be used. For such techniques to work in high dimensions, if the number of features is d and the number of training samples is N, then we should have $N \gg 2^d$, which is practically infeasible for large d. This is known as the curse of dimensionality in ML.

8.4.2   Naive Bayes Classifier

A naive Bayes classifier assumes the conditional independence of features and uses the Bayes' theorem to predict the posterior probability of a class given the observed data sample. Bayes' theorem finds the most likely "cause," which led to the observed "effect." In classification, the effect is the data sample and its features, and the cause is its unknown class. The posterior probability of a cause is proportional to the product of the prior probability of the cause and the probability of the effect given the cause. Training data allows you to estimate both of these values. The approach is called *naive* because we treat each feature independently, when in reality there may be dependence between features.

8.4.3   FDA, LDA, and QDA

Fisher Discriminant Analysis (FDA) deserves a special mention, being one of the first techniques in classical ML. FDA is easy to understand in the 2-D case for two classes. Find a direction such that when data samples are projected on it, the dual goal of making the interclass separation high and making the intra-class variation low is met. This is achieved by maximizing the Fisher criterion,

$$\frac{(m_1 - m_2)^2}{(\sigma_1^2 + \sigma_2^2)},$$

where m_1, m_2 are the means of the *projected* 1-D classes and σ_1, σ_2 are their standard deviations. There is an analytical solution based on linear algebra. Given the covariance matrices and the means of the given 2-D classes, the direction which maximizes the Fisher criterion is

$$w = (\Sigma_1 + \Sigma_2)^{-1}(\mu_2 - \mu_1).$$

By assuming that the features of the two classes follow a Gaussian distribution, with both distributions having the same variance, and by applying the Bayes classifier, we get Linear Discriminant Analysis (LDA). We get Quadratic Discriminant Analysis (QDA) if we drop the homoscedasticity assumption of equal variances.

8.4.4 Support Vector Machines

An SVM classifier, in its linear form, looks for the maximum margin between two classes in the feature space, in the middle of which it can place the separating hyperplane. The effectiveness of an SVM comes in its nonlinear form, where it transforms the features to a high-dimensional space using the kernel trick. One never really goes to the high-dimensional space. The kernel trick is employed to efficiently compute inner products in the high-dimensional space as a shortcut method. The equivalent separating boundary in the original, lower-dimensional feature space can then be highly nonlinear. The slack hyperparameter allows an SVM to find a soft margin with allowance of a limited amount of violations of the margin constraint.

8.4.5 Neural Networks

Classical neural networks work in the feature space. They have one hidden layer and are considered shallow networks. Classical neural networks can be viewed as a network of logistic regression modules because the sigmoid activation function is used for every neuron, unlike modern AI models which use ReLU. Another conventional choice of the activation function is the hyberbolic tangent function.

The *universal approximation theorem* shows that the classical neural networks can approximate any continuous function on a compact set (bounded and closed set). The set of neural network functions is dense in this target space. The theorem is of theoretical interest without practical implications, because it shows that, by increasing the number of hidden units, we can approximate a function in a piecewise manner with commonly used activation functions. The requirement is that the activation function should be nonconstant, bounded, and continuous. An example is the sigmoid function. Note that the sum function $\sigma(x) + \sigma(-x)$, where σ is the sigmoid function, will create a building block for any smooth function. Universal approximation is like building a house brick by brick, where bricks are derived from such a bumpy activation function.

It can be shown that the classical neural networks can implement arbitrary Boolean functions. At the input layer, for a True variable, use +1 value, and for a False variable, -1 value. On edges, use +1 connection weight to check whether a neuron is true, and use -1 for false, and choose appropriate bias terms to implement the Boolean gates.

In the worst case, the number of neurons in a shallow network is exponential in the input dimensionality, and, in the exact case, in the number of statistically independent features.

8.4.6 Decision Trees

Classification and Regression Trees (CARTs) have been influential due to their speed of inference and the simplicity of the approach. They are more interpretable when they are small. CARTs are generalized to random forests to make them more robust and to reduce their variance. In random forests, multiple Classification and Regression Trees (CART)s are trained by using randomized subsets of the training data and by using randomized subsets of the features considered at each tree node. This is an ensemble approach, and, coupled with the above randomization trick, random forests can work well in practice. Random forests have played a big role in the adoption of ML by industry.

8.4.7 Gaussian Process Regression

Gaussian Process Regression (GPR), also known as Kriging, models the observations as samples of a multivariate Gaussian distribution whose parameters are learned. If we have seen N input-output (x_i, y_i) pairs, we model the observed function values as a sample vector from an N-dimensional normal distribution,

$$\mathcal{N}(Y_1, Y_2, \ldots, Y_N),$$

where the covariance of Y_i and Y_j depends on the observed x_i and x_j. To predict the $(N + 1)$-th value, the Gaussian function is extended to $(N + 1)$ dimensions,

$$\mathcal{N}(Y_1, Y_2, \ldots, Y_N, Y^*),$$

under the same covariance structure of values. The conditional distribution of the new dimension $Y_{N+1} = Y^*$ on the observed samples is used to predict the function value y^* at the new input x^*.

Suppose you are given just one value $y = 5$ at $x = 1$. What is the prediction for y at $x = 2$? Assume a covariance structure, which tells how much a sample at x will affect a sample at $x + \Delta$. It is given by the covariance function $k(x, x')$ known as the kernel of the Gaussian process, and it measures the dependence between the function values at different inputs x and x'. The kernel is chosen according to the application. Suppose it is a kernel which correlates nearby points with small Δ strongly. If Δ is large, then they correlate less strongly. Thus, assuming a 2-D Gaussian distribution with this covariance structure, one can predict y at $x = 2$ by conditioning Y_2 on the previous observed value $y_1 = 5$ at $x = 1$, that is, by taking a cross section through the 2-D Gaussian at $y_1 = 5$. The mean and the variance of the conditional distribution $P(y_2|y_1 = 5)$ give the predicted value and the uncertainty around it.

Let us say we observe the value y at $x = 2$. The kernel function has hyperparameters, which are fine-tuned using either full Bayesian inference or by maximization of the likelihood of the observations. To predict the value of y at $x = 3$, we repeat the same but with covariance influence of the sample at $x = 1$ less than that of the sample at $x = 2$ on the new sample at $x = 3$. For each prediction, GPR outputs the uncertainty interval around it. This uncertainty reduces with an increasing number of observations.

An application of GPR is to model seasonal and periodic processes. For example, for a business, sales today will correlate with those of last week and of last year, on similar days and holidays. One has to choose a suitable kernel function to model such periodic correlation.

The GPR method is also used for the problem of hyperparameter search in AI. Given the performance of an AI model on certain values of training hyperparameters such as the learning rate and the momentum factor, GPR predicts the performance on new hyperparameters.

8.4.8   Unsupervised Methods

We mention that k-means clustering is a well-known unsupervised ML technique to explore the geometry of data in terms of naturally occurring clusters. Vector quantization (VQ) is a general term for clustering, which includes the k-means algorithm.

The expectation-maximization (EM) algorithm is a general formulation to find the model parameters in the presence of unseen latent variables or incomplete data. In particular, the EM algorithm applied to model the observed data as a mixture of Gaussians is a soft variation of k-means. The latent variable for an observed sample here is the information about which Gaussian function generated the sample.

Density-Based Spatial Clustering of Applications with Noise (DBSCAN) is another popular clustering technique due to its robustness to outliers and ability to find arbitrary-shaped clusters.

Principal Component Analysis (PCA), which is an application of Singular-Value Decomposition (SVD) of matrices, has been used for dimensionality reduction, exploratory visualization, and latent factor discovery for decades. A supervised version of dimensionality reduction called Partial Least Squares (PLS) has been applied, where we have a large number of features compared to the number of observations. In multivariate statistics, Multidimensional Scaling (MDS) has been used for visualization of high-dimensional data in a low-dimensional space. A modern method for such visualization is t-Stochastic Neighbor Embedding (t-SNE). Kohonen self-organizing feature maps (SOFM) and competitive learning have been used to visualize the geometry of data, and they can be used to discover clusters as well.

** Exercise 22   Classical ML

1. State for which of the following ML models the loss function is convex: linear regression, logistic regression, SVM, or neural networks?
2. What is the hyperparameter used in SVM? What is the name of the loss function which SVM uses?
3. During training of a classification tree, what should be the criterion to decide to split a node or not?

8.5   XGBoost

We devote this full section to Extreme Gradient Boosting (XGBoost) because of its great success in practice, see [30]. XGBoost is a popular implementation of the Gradient Boosted Machine (GBM).

It starts with the idea of boosting, which is that of cascading many classifiers to build a strong one, in which the subsequent classifier boosts the score output by getting trained on examples, which are difficult for preceding classifiers. Boosting is an ensemble method, which can turn a collection of weak learners into a strong one.

Consider the regression problem under the L^2 loss. In GBM, boosting stages are implemented by decision trees. The scores of all trees get added up to compute the final score. The training happens as follows. The first tree is fitted to the function. There is an error or residual. A second tree is fitted to this error, and the process is repeated till the desired accuracy is achieved without overfitting. It is like taking gradient steps in the minimization of the error. See Fig. 8.2. For a sample x, let G be the ground truth and S be the sum of scores (predictions) output by the first k trees

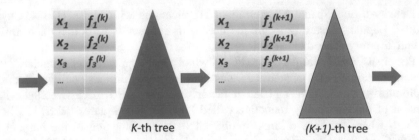

Fig. 8.2 XGBoost algorithm, which is based on gradient boosted trees. A boosting tree is added to fit to the residual error for each data sample. For the L^2 regression, it is the residual error $(G_i - y_i)$, where y_i is the fitted value by the preceding trees for the sample x_i and G_i is the ground truth. In general, for a differentiable loss function, the gradients have to be computed. These gradients are shown as $f^{(k)}$ for the k-th boosting stage. The k-th tree is fitted to the gradients, lowering the loss further. The process is repeated at the next stage. For quadratic approximation, a second-derivative term is added

which have been fitted. Suppose we are building the $(k + 1)$-th tree, which outputs score s for the input x. Therefore, we want to minimize

$$(G - (S + s))^2.$$

This is the same as doing regression on the residual by minimizing

$$((G - S) - s)^2.$$

Note that since G and S are fixed and s is variable when we are building the $(k+1)$-th tree, the above reduces to the minimization of

$$2(S - G)s + s^2 + \text{Constant Term},$$

where the constant term is $(G - S)^2$. The above idea of fitting to the residual error is mathematically correct for MSE-based regression, and the above expression points to the technique for arbitrary differentiable loss functions. In general, we may not know the ground truth; therefore, we can't compute the residual. For example, consider the problem of ranking of retrieved documents in Web search. We know the relative scoring but not the absolute scores. Still we can apply XGBoost as follows. We just rewrite the above expression in the Taylor series form,

$$L(S + s) - L(S) = L'(S)s + \frac{1}{2}L''(S)s^2,$$

where L is the loss function for the input x as a function of its score. Note that the regression loss function discussed above is $L(S) = (G - S)^2$. Therefore, $L'(S) = 2(S - G)$ is the first derivative of the loss function $L(S)$ with respect to S, and

$L''(S) = 2$ is its second derivative. The gradient for the current boosting stage is computed with respect to the score fitted by the previous boosting stages.

In general for a differentiable loss function, a function which computes the gradient of the loss function for a data sample x is needed. For quadratic approximation, the second-derivative term is added. A regularization term is added to the loss function, which is a combination of penalty terms for the number of leaf nodes and for the magnitude of predictions at the leaf nodes.

As we are building a new tree, we make the decision to split a tree node or not by looking at whether it lowers the regularized loss function. This is done by computing the gain

$$g = C_N - (C_R + C_L),$$

where C_N is the sum of the loss function values at samples landing in the tree node N and C_L and C_R are the sums of the loss function values at samples landing in the proposed left child and right child, respectively, in the proposed split. If g is high, then we go for the split, because the split lowers the overall loss.

> *Classical ML is rich in a wide variety of techniques, which are available to practitioners who want to work in the feature space. Gradient boosted trees and random forests have proven to be widely useful in practice.*

8.5.1 Relevance Ranking

The effectiveness of XGBoost lies in its ability to handle any loss function. What if we can't directly compute the gradient of the overall objective function as in ranking, but we do know about pairwise relative ranking? For Web search, we need to rank retrieved Web documents. The loss is now based on pairwise document-ranking scores weighted by a relevance-ranking metric such as NDCG. Here is the high-level intuition how to adapt gradient boosted trees for this task.

LambdaMART based on Multiple Additive Regression Trees (MART) uses gradient boosting to rank documents; see [22]. Lambda (λ) refers to the gradient, a force on each document to change its search position upward or downward. It is the residual error on ranking done by the trees built so far. The total force is aggregated from all pairwise tugs a document experiences in the pairwise comparisons.

Suppose k trees built so far output some relevance scores for the retrieved documents. We perform pairwise comparison of documents to compute λ, that is, the derivative of a suitably defined loss function, which takes the difference in the two scores as the input. The score of the higher-ranked document should be larger than that of the lower-ranked document, resulting in low loss. Then, λ points in the direction in which the pairwise score difference has to change.

Modulate λ by the expected NDCG gain by swapping the document pair, so that there is greater force on documents where we need to fix the residual error for larger gains in NDCG. For example, suppose a highly relevant document A is currently ranked very low compared to an irrelevant document B, and therefore, NDCG will benefit a lot if A and B swap positions. By multiplicatively scaling $\lambda_{A,B}$ with the pairwise NDCG gain, we will make the algorithm A move up more aggressively, and B will be moved downward. Add up the pairwise forces on the document A obtained by pairwise comparisons with other documents to compute the total λ_A.

Next, $(k + 1)$-th tree takes a boosting step and performs regression on the computed λ values of documents. The magnitude of the gradient step is controlled by a learning rate hyperparameter. This adjusts the scores in the direction of overall NDCG gain.

8.6   Chapter Summary

Classical ML consists of several techniques and derives inspiration from statistics, estimation theory, optimization theory, and Bayesian inference; see [16, 51, 103]. Classical ML works in the feature space rather than the raw data space. Maximum likelihood estimation of parameters maximizes the likelihood of observed data. Bayesian inference incorporates prior beliefs in the estimation theory, and its effectiveness derives from iterative update of beliefs on presentation on new evidence. Linear models with regularization are widely studied techniques and offer intuitive geometrical interpretations of their working. A number of supervised and unsupervised techniques have been developed. For supervised techniques, random forests and gradient boosted trees have been influential in practice and have been widely adopted by industry. For unsupervised techniques, PCA and the k-means algorithm are two well-known examples. Classical ML systems are capable of solving practical problems, for example, relevance ranking in Web search can be solved by gradient boosted trees with a customized ranking-based loss function.

Problem Set 8

1. In regression, what will be the disadvantage of choosing the L^p loss for some $p \geq 3$? What will happen as $p \to \infty$?
2. Prove that the mean minimizes the sum of the squared deviations

$$\sum_{i=1}^{n}(a_i - x)^2$$

   of data samples $a_1, \ldots, a_n$.
3. How does increasing the value of k affect the bias and the variance of a k-nearest neighbor model?

4. How does increasing the value of the slack parameter affect the bias and the variance of an SVM model?

5. What is a nonparametric ML technique? Give an example. What is a parametric ML technique? Give an example.

6. Why is taking an ensemble of a number of ML models often beneficial? The outputs of different ML models are averaged in an ensemble.

7. Provide two related closed-form solutions for linear regression.

8. What is one limitation of Gaussian process regression?

9. Suppose you have a prior belief of 0.9 that your friend is happy. You notice that when the friend is happy, then the chance of going for bowling is 0.95. When the friend is not happy, the chance is only 0.01. You do go for bowling. How will you revise your belief?

10. The way SVMs work is reminiscent of the way a neuron in an AI model cuts a feature space with a linear function, which corresponds to a nonlinear boundary in the input space. Why do AI models outperform SVMs? List all the differences between the two approaches.

Part II
Applications

Deep Learning is a superpower. With it
you can make a computer see,
synthesize novel art, translate
languages, render a medical diagnosis,
or build pieces of a car that can drive
itself. If that isn't a superpower, I don't
know what is.

Andrew Ng

Chapter 9
To See Is to Believe

The eyes are more exact witnesses
than the ears.

―――――――――――――――――――

Heraclitus

You can't depend on your eyes when
your imagination is out of focus.

―――――――――――――――――――

Mark Twain

It has been said that humans learned to stand and walk upright in order to see things faraway in the plains of Africa. We are blessed with an amazing sense of sight which allows us to perceive what is happening around us. As all new parents know, a newborn baby starts noticing them and the world. We can segment out objects in our visual field, recognize them, and infer their attributes. We can detect motion and are able to predict what is likely going to happen next.

AI scientists and engineers have made significant advances in making machines which do the same. Computer vision has been an active area of research for decades. In this century, the field got tightly coupled with that of ML and recently with AI. There are several problems of interest in computer vision and the input can come in different forms. The problems look different, but they are all interrelated and can be considered to be special cases of an active process of scene understanding. The input can be images, videos, 3-D volumetric data, or 3-D point clouds. Let's look at some applications in the visual domain. See Fig. 9.1.

S. Dube, *An Intuitive Exploration of Artificial Intelligence*,
https://doi.org/10.1007/978-3-030-68624-6_9

Fig. 9.1 The reference image is at the top left. The common computer vision problems of image similarity, object detection, and semantic segmentation are shown at the top right, bottom left, and bottom right, respectively. Semantic segmentation shown here is instance aware. Its simpler variation is instance agnostic. Image classification results for the two top-row images should be "flower," "leaves," and "plant" with high probability

9.1   Image Classification

9.1.1   Motivating Examples

Given a photograph, video, or 3-D data, we want to classify it. Some examples:

1. What handwritten digit, among 0–9, is it?
2. What species of bird is it?
3. What species of plant is it?
4. Is this satellite picture indicating early signs of drought?
5. Is this satellite picture indicating deforestation?
6. Is this aerial drone picture of ocean water indicating eutrophication?
7. Does this X-ray picture indicate a healthy normal state?
8. Is this picture of objectionable content not suitable for children?
9. Is it a picture of a human face?
10. What breed of dog is this?

11. What model of car is this?
12. What type of furniture is this?
13. What kind of dress is this?
14. What type of shoe is this?
15. What word is spoken in sign language in this video?
16. Is this video of a basketball game?
17. Is this video of a baseball game being played by children?
18. Is it a burglar in this surveillance video clip?
19. Is it a dog in this surveillance video clip?
20. Is a person running in this surveillance video clip?
21. Does this ultrasound clip indicate an abnormality?
22. Is this a wedding video?
23. Is this photo of bird that of puna yellow finch or of greater yellow finch?
24. Is this 3-D point cloud that of a car or a truck?
25. Does this sub-volume of a 3-D CT scan have a region of interest?
26. Is it a traffic intersection?

In several examples above, an implicit assumption was made that the object of interest dominates the picture when it is present, and if background is there, then it is not excessive. If, in a large picture, a cat occupies a handful of pixels surrounded by clutter and several objects, then it will become a difficult problem to classify the image as that of a cat. A more appropriate model will be that of object detection, which detects (localizes) all the objects in the image and can be viewed as multi-label classification with localization.

The yellow finch question in the above list is that of fine-grained classification, in which we push the limits of state-of-the-art classification further. The goal is to discriminate objects based on subtle and fine features, and examples are classification of species of related birds, flowers, insects, or models of vehicles.

9.1.2 LeNet

The handwritten-digit recognition problem led to one of the first CNNs in the world in the 1990s. Designed by Yann LeCun et al. in 1998, based on research dating back to 1988, and referred to as LeNet, this CNN is viewed as the first major milestone in AI-based visual perception. These days, the first problem which AI students experiment with is MNIST digit recognition, where MNIST is the name of the dataset. See [123, 124] for this seminal work in convolutional neural networks trained by the backpropagation algorithm for the first time in the world.

> *LeNet for the MNIST dataset is the first major milestone reached in the 1990s in CNN research and would later usher in an era of AI-based computer vision in the second decade of the twenty-first century.*

9.1.3   Stacked Autoencoders

In the middle of the first decade of the twenty-first century, a series of influential papers were published which revived interest in multi-layer neural networks; see [15, 88, 90]. These papers introduced a novel approach of training deep neural networks by pre-training their hidden layers, one after another, using unsupervised techniques, for restricted Boltzmann machines and then for autoencoders.

Unsupervised pre-training was followed by a supervised fine-tuning stage for classification tasks. This was a breakthrough because the authors were able to train a novel multi-layer neural network and it reignited interest in neural networks. Though, in a matter of a few years, supervised techniques for deep neural networks were soon to take over the world, the work done in 2006 is considered a turning point in AI research.

9.1.4   AlexNet

A third breakthrough occurred in 2012 when a CNN-based AI model won the ImageNet competition. Supervised learning was employed to train a CNN with a large number of neurons and several hidden layers on a large training dataset. Influential techniques such as ReLU [59] and dropout were introduced which greatly helped the training process; see [116]. Data augmentation was employed to expand the training data by adding images obtained by applying transformations to the training images. The neural network was called AlexNet, named after the first author, a graduate student of Geoffrey Hinton.

The work stimulated development of new models, which were submitted to the ImageNet competition in subsequent years, and several noteworthy neural networks came out of this effort. The architecture of AlexNet, which consisted of several convolutional layers, some of these layers immediately being followed by a max pooling layer, and the convolutional layers being followed by a few fully connected layers just before the softmax layer, had influence on the architectures which were developed soon after. Note that fully connected layers are a special case of convolutional layers in which the size of the input blob and the filter are the same. The names of AlexNet layers such as Conv5 and FC6 became well known in the AI community.

> The annual ImageNet competition provided a strong stimulus to AI-based computer vision research starting in 2012.

9.1.5 VGG

An ImageNet competition model is VGG (2014); see [200], which exists in two flavors, VGG-16 and VGG-19, the numbers indicating the number of layers. VGG stands for the Visual Geometry Group of Oxford University. VGG models used strictly 3×3 filters, because stacking layers of small filters makes their receptive fields as big as larger filters with the added advantage of increasing the expressive power of CNNs. Three layers of 3×3 will have the same receptive field as a 7×7 filter, and at the same time, the network will be deeper. Though the VGG models are large in the number of parameters, they are widely used for feature extraction in computer vision tasks. We will discuss the application of neural networks as feature extractors later in this chapter when we talk about the image similarity problem.

9.1.6 ResNet

An influential ImageNet competition model was developed by Microsoft Research in 2015. It is called ResNet, which stands for residual network; see [79]. The inspiring goal was to increase the number of layers in a CNN to hundreds and even as high as one thousand, and still be able to train them without having the problem of vanishing gradients. The goal was achieved with the brilliant idea of using skip connections. Till 2015, DNN followed a layer-wise cascaded structure. With skip connection, computation flow can be split into two paths: (1) normal cascade of layers and (2) a skip path which bypasses these layers. The two merge back later and the data flowing along these two paths get added up. As an analogy, (1) is like going through the traffic lights of a city, and (2) is like taking a bypass freeway, both eventually merging. The computation graph starting from the split to the merging step is called the residual block, and ResNet consists of a cascade of many such blocks. Therefore a residual block computes:

$$G(x) = F(x) + x,$$

where F is computed along path (1). In other words

$$F(x) = G(x) - x$$

is the residual.

To see why this is a powerful idea, suppose we have built a network with k residual blocks already and we want to add one more residual block to continue the cascade. The k existing blocks define a manifold or a function. If this manifold or function captures the complexity of the problem, then the $(k + 1)$-th block is unnecessary and therefore its $F(x)$ will be zero, and the data will just flow along the skip path. In case the manifold or function needs additional fix-ups, then F

will encode these fix-ups, which are residuals. Therefore with more residual blocks, finer details can be added. Conceptually the idea is similar to gradient boosting in classical ML.

> *ResNet carves out a manifold or function in such a way that finer details are added to the carved manifold or the fitted function with the addition of more residual blocks.*

9.1.7   Inception V3

Another ImageNet competition model is Inception V3 (2016), the result of work by Google engineers; see [211]. Despite its colorful name inspired by the Inception movie (2010) where dreams occur within dreams, the network is more prosaic consisting of cascaded inception modules, in each of which convolution filters of different sizes are applied on the same input. The output feature maps are concatenated. Batch normalization is used throughout the model and applied to the activations of the neurons before ReLU is applied. Dropout is used at the final fully connected layer.

Filters are small in size, namely, 1×1 and 3×3. Note that the convolutions are 3-D so 1×1 is really $1 \times 1 \times F$, where F is the number of feature maps in the input blob. Since a $1 \times 1 \times F$ filter is equivalent to a layer of FFN with F input neurons, therefore it is like a one-layer *FFN* embedded inside an Inception module. It is a network-in-network structure; in the Inception movie it is dream-in-dream.

One way to understand Inception V3 is to think of it as a feature extractor. It takes an input image of dimensions $229 \times 229 \times 3$ (an RGB image) and produces an output feature blob of dimensions $8 \times 8 \times 2,048$. The output feature blob is fed into a fully connected layer to produce the final classification output, or it can be used as a representation of the input image for computer vision problems such as image similarity and image captioning.

9.1.8   Showroom of Models

Several CNN models have been developed for image classification problems by the AI community. Some of these are intended to be lightweight and have been motivated by applications at-the-edge devices with limited compute power. See [233, 237, 238] to learn about a wide selection of models. Let's take a 10,000-ft.-high view of the world of CNN models:

- ZFNet (2013) makes minor modifications to AlexNet, and it won the ImageNet 2013 competition [251].

- GoogLeNet (2014) introduces the idea of Inception modules. It is also known as Inception V1 [208].
- Inception-ResNet V2 (2016) applies the skip connection idea from RestNet to Inception V3 [209].
- Xception (2016), which stands for "Extreme Inception" replaces Inception modules with pointwise 1×1 convolutions followed by depth-wise 2-D convolutions applied independently to channels [36].
- DenseNet (2016) uses skip connections, and unlike ResNet, which performs addition, it performs concatenation [96].
- SqueezeNet (2016) is lightweight, consists of fire modules consisting of squeeze and expand layers, and seeks to achieve the same accuracy as AlexNet [98].
- ResNeXt (2016) generalizes ResNet by replacing the computation subgraph inside the residual block of ResNet with an additive homogenous multibranch graph [246].
- ShuffleNet (2017) divides the convolution filters into groups to optimize the number of parameters and shuffles them to improve the information flow between these [254].
- NASNet (2017) applies Neural Architecture Search (NAS) to discover networks automatically; NASNet-Large and NASNet-Mobile are two outcomes; see [258].
- MobileNet V2 (2018) is a lightweight network which uses depth-wise separable layers (depth-wise 2-D convolutions followed by pointwise convolutions) and skip connections between linear bottleneck layers (thin layers with a compressed number of channels) [186].
- DarkNet-53 (2018) is used by the object detection algorithm YOLO-v3 and can be used for image classification [175].
- EfficientNet (2019) scales up a baseline model in width, depth, and input resolution yielding a family of CNNs from EfficientNet-B0 through EfficientNet-B7 [212].
- AmoebaNet (2019) is obtained by applying an evolutionary algorithm to search for the best neural architecture [174].
- SENet (Squeeze and excitation) (2019) squeezes global information in channels through an aggregate operation and then pushes those channels through a mapping to compute a channel-wise attention weight vector to scale feature maps [95].

The success of different models in achieving their goals indicates that fundamentally there are many possible variations of AI models which can perform high-dimensional sculpture of manifolds with varying degrees of approximation accuracy, and representation learning doesn't have a unique solution.

** Exercise 23   CNN Parameters

Compute the number of parameters in MobileNet-V2, ResNet-50, ResNet-152, Inception V3, VGG-19, and NASNet-Large. What will be the model size in MB?

9.1.9   Your Own Network

Suppose you need to solve an image classification problem. Your best bet is to choose a pre-trained AI model from the public domain and fine-tune it for your application. Fine-tuning or transfer learning means freezing most of the lower layers and allowing backpropagation to tweak a handful of higher layers on your own data. This is an effective approach to train customized computer vision models with a limited amount of data, and in NLU, it has been recently making an impact; see [48]. Another option is to build one's own CNN and train it from scratch, provided there is enough training data. You can also apply NAS, which typically uses either evolutionary, differentiable, or RL techniques. Evolutionary algorithms spawn, grow, combine, mutate, and prune candidate models. An edge in the computation graph can be taken as a differentiable weighted combination of candidate operations, which can be optimized. Finally, RL can be applied by sampling architectural choices from the output of a recurrent policy network, with the reward being the accuracy of the sampled model; see [258].

Once the network is trained or fine-tuned, its performance should be evaluated on test data with metrics such as confusion matrix, accuracy, ROC curve, AUC, mAP, PR curve, and F1 score. Since it is becoming relatively easy to train CNNs for image classification tasks, thanks to a commendable community effort, any setbacks one may face in solving practical computer vision problems are likely to be temporary.

Using a pre-trained CNN model and fine-tuning it with task-specific training data is an effective technique to solve several computer vision problems.

9.2   Object Detection as Classification

Object detection is finding where the object is in a given image. There are three major approaches; see Fig. 9.2:

1. Sliding window-based classification approach.

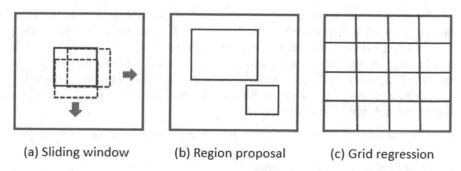

(a) Sliding window   (b) Region proposal   (c) Grid regression

Fig. 9.2 Object detection is classification with localization of the object. There are three primary approaches for object detection: (**a**) sliding window, (**b**) region proposal, and (**c**) grid regression. The sliding window approach exhaustively tries every possible position. The region proposal method looks at only those regions which are likely to have the object. One doesn't have to run the CNN separately in each region, because the image features computed by a backbone CNN on the whole image can be pooled for each region. The grid-based method formulates the detection as a regression problem for each grid cell. It predicts whether the object is centered inside a given cell, and if it is centered, then its bounding box is predicted (the bounding box need not be inside the cell)

2. Region proposal-based classification approach.
3. Grid-based regression approach.

It is a harder problem than classification because the output is classification with localization information. The good news is that we can trivially reduce object detection to classification. We will look at the sliding window approach in Sect. 9.2.1 and the region proposal approach in Sect. 9.2.2. In detail, we will discuss the grid-based methods in Sect. 9.3.2. Before we develop intuition about these different methods, let us look at how we will evaluate the results of object detection.

For an object detection task, the ground-truth set consists of correct bounding boxes with classification labels. Object detection is a retrieval problem, and the bounding boxes which are returned by the AI model are looked at for correctness. To check whether an output bounding box is the same as a ground-truth one, their overlapping area is measured by Intersection Over Union (IOU). IOU is the Jaccard coefficient with pixels as elements of the two bounding box sets, and it is the ratio of the overlapping area to the total area of the union of the two boxes. Incorrectly retrieved boxes (with small IOU) are false positives, and correctly retrieved ones (with large IOU) are true positives. The mAP metric, which is the area under the PR curve, is used to evaluate the results.

---

* Exercise 24   PR vs ROC

Why do we use the PR curve, instead of the ROC curve, for object detection?

One has to quote mAP numbers with respect to an IOU threshold. Therefore one states mAP@0.5 meaning IOU has to be over 0.5. Sometimes mAP@[0.5,0.95] with step size 0.05 is used. This means that one takes the average of different mAPs obtained by varying IOU from 0.5 to 0.95 with a step size of 0.05. State-of-the-art mAP@0.5 numbers for different techniques on the MS-COCO dataset [133] are in the 40s through 60s (in percentage). Numbers for mAP@[0.5,0.95] are currently in the 20s through 40s (in percentage).

9.2.1   Sliding Window Method

To find dogs in an image, train a dog classifier on tightly cropped dog images which all have been resized to the same size $w \times h$. For a given image of size $W \times H$ where we assume $W \geq w$ and $H \geq h$, we apply the classifier on all of its regions, small or big, after resizing them to $w \times h$. It is a brute-force exhaustive method. The same principle will solve the game of finding Wally (or Waldo) in the well-known book series *Where's Wally*. Just pass a sliding window roughly the size of Wally across the image. Therefore, the approach is

$$y = F(\text{ROI}(x)),$$

where ROI is a function which returns all possible regions of interest inside the image x and F is the classifier. We get an output for each ROI, and therefore, a post-processing stage is needed to produce the aggregated output. Often, we add a regression head R which corrects or fine-tunes the best ROI region to produce the final bounding box coordinates B:

$$B = R(\text{ROI}(x)),$$

where R is constructed by adding a few layers on top of the same backbone CNN which F uses. If R is omitted, we still have the location of the object in the form of ROI. A key observation is that the whole image, which is bigger than $w \times h$, can be input to a modified $W \times H$ classifier obtained from the trained $w \times h$ classifier by enlarging the width of the convolutional layers. The output now is a $p \times q$ classification map rather than a single number. It is a heat map or a class activation map. Hot values in this output are clustered together, and the modes of these clusters give the locations of dogs in the image. Alternatively each hot value is centered on a $w \times h$ bounding box around a dog. In order to fuse overlapping bounding boxes, Non-Maximum Suppression (NMS) is employed. In NMS, bounding boxes which overlap significantly with a maximum-scoring bounding box are deleted, and the process is repeated till boxes can't be pruned further.

9.2.2   *Region Proposal Method*

Instead of applying the classifier exhaustively, a natural idea is to apply it in only those regions which are likely to have an object, and this idea leads to Region-based CNN (R-CNN). Therefore, we make the ROI function above smarter. The dog classifier is applied on regions which are proposed based on heuristics or another ML system, referred to as the selective search method. They proposed regions get resized to the same size $w \times h$. The same goes for detection of Wally, who wears a red stripy T-shirt in the *Where's Wally* books. Use some quick way of finding red stripy regions as potential candidate regions, and apply a classifier on each. A regression head is often added to the same backbone CNN to fine-tune the final bounding box coordinates.

The above idea behind R-CNN has been extended significantly further. The region proposal component is folded into the same classifier network, leading to significant speedups. The key observation is that one does not have to apply CNN separately to each of the regions, because the shared convolutional features need to be computed only once, and one can just extract or pool the proposed regions from the same feature blob directly. Furthermore, the regions can be proposed based on a joint learned network which works on the feature blob rather than the input image. Therefore we have the following:

$$z = F(x),$$

$$y = G(\text{RPN}(z), z),$$

where z is the feature blob computed by a backbone CNN F, RPN is the region proposal network, and G is ROI pooling followed by the classification network. In addition, a fine-tuning regression head is also attached as in the case of the sliding window approach. Note that a number of region proposals are returned by the RPN network, which all have to be passed through G, so it is a two-stage approach. These advanced models are known to be faster variants of R-CNN; see [178] for all the fine details which have to be worked out to design such an integrated system.

See Sect. 9.3.2 for a discussion on how the bounding box coordinates are predicted explicitly using regression. In the region proposal methods as well as the sliding window method, regression is still employed because the bounding box coordinates are fine-tuned by attaching additional layers which perform regression to adjust the coordinates finely.

9.3   Regression on Images

9.3.1   Motivating Examples

Consider the following examples:

1. Given an image of a cat, output the bounding box giving the location and the size of the region containing the cat.
2. Given a pair of consecutive video frames, compute a dense motion field.
3. Given an image, output a quality number indicating how well focused the image is.
4. Given an image, output a quality number indicating how aesthetically appealing the image is.
5. Given a drone picture, count the number of cars in the parking lots.
6. Given a medical image, output the level of disease progression.
7. Given a movie, predict the age of the audience for which it is suitable.

For simple regression tasks, change is minor, and the model can be trained under a regression loss such as L^2 error. For some tasks, a larger system must be broken into components. For example, the preferred way to count the number of cars will be by detecting cars in the picture using object detection and counting the number of detected cars.

9.3.2   Object Detection as Regression

Object detection involves both classification and regression. Detection of dogs and cats involves finding the bounding boxes of dogs and cats in the image and the classification labels of these boxes. This is a special case of attribute learning where the bounding box coordinates are numerical and the label is categorical.

Typically two heads are attached to a backbone CNN which outputs a blob of features. The classification head attaches additional layers to the base feature extractor network for classification, and the regression head attaches its own separate layers for regression. Recall that the regression head can be also used to fine-tune the bounding box coordinates in the methods discussed in Sect. 9.2.

Let's build an intuition into regression-based methods which output all the bounding boxes in a single shot. This idea has led to the development of the You Only Look Once (YOLO) [176] and Single-Shot Detector (SSD) [135] family of detectors. These are single-stage or single-shot methods which contrast with two-stage methods such as R-CNN.

9.3.2.1 Regression Output

Suppose we are given $N \times M \times 3$ color images and the goal is to detect objects in C positive classes. For each pixel, define the following attributes (assume that there can be at most one object centered on a pixel):

1. Negative Class Label z. This label is 1 if and only if the pixel is negative class, that is, none of the C-positive classes have any object centered at it. Semantically, it is a background pixel or a pixel belonging to an uninteresting class.
2. Positive Class Label Vector $L = (l_1, l_2, \ldots, l_C)$, where $l_j = 1$ if and only if an object of class j is centered at the pixel. There can be at most one l_j which is 1. If none is 1, then z has to be 1.
3. Bounding Box Attributes (w, h), the width and the height of the bounding box containing an object in class j when $l_j = 1$.

Note that if $z = 1$, then the rest of the attributes are non-applicable. Therefore, there is a $C + 3$-dimensional attribute vector associated with each pixel. For the whole image, we will have a ground-truth attribute blob of dimensions:

$$N \times M \times (C + 3).$$

Call this blob B. An *object detection* CNN converts an $N \times M \times 3$ input into an $N \times M \times (C + 3)$ output:

$$B = F(I)$$

where I is the input RGB image and F is the function implemented by the CNN.

9.3.2.2 Grid-Based Approach

Object detection is made computationally more efficient by working with a coarse grid. Divide the image into cells (nonoverlapping tiles) to give an $n \times m$ grid. Now the dimensions of B reduce to the following:

$$n \times m \times (C + 5).$$

Note that there are $C + 5$ attributes of a grid cell:

1. Negative Class Label z. Same as before.
2. Positive Class Label Vector L. Same as before.
3. Bounding Box Attributes (c_x, c_y, w, h), the center, the width, and the height of the bounding box, respectively. These numbers are often normalized with respect to the grid cell size. Note that an object doesn't have to be totally inside a grid cell. The object just has to be centered in the cell and can extend outside the cell.

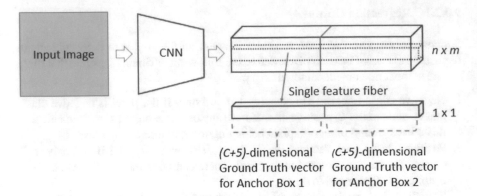

Fig. 9.3 General framework to solve computer vision problems with CNN. If $n = m = 1$ with only the classification ground truth, we have the standard image classification. The grid-based approach in which there are multiple anchor boxes serves as the basis for most practical object detection algorithms. The ground truth is a mix of categorical variables and regression variables, and therefore the loss function is a combination of the classification and regression loss functions. For finer grid at the pixel resolution, we solve the problem of semantic image segmentation

We now relax the constraint that there can be at most one object inside a grid cell. Let M denote the maximum number of objects centered in the grid cell. See Fig. 9.3, in which $M = 2$.

The size of an attribute vector for a grid cell with multiple objects is

$$n \times m \times (C + 5) \times M.$$

See Fig. 9.3 to confirm the above count. Suppose $M = 5$, and we have two dogs, one cat, and one child centered in a grid cell. What should be the order of the objects in the attribute vector? Should it be

Dog, Negative, Dog, Cat, Child?

Or some other one, such as

Child, Cat, Negative, Dog, Dog?

In order to define the ground truth unambiguously, geometrical semantics are added to each position. For each grid cell, five pre-defined *anchor boxes* of different sizes and shapes are defined:

$$A_1, A_2, A_3, A_4, A_5,$$

where A_i is associated with position j in the ground-truth vector. An object is assigned to position $k \in \{1, 2, 3, 4, 5\}$ if its bounding box matches the corresponding A_k in terms of IOU. Therefore, if the child is geometrically consistent with the size and shape of A_3, then it will occupy the third position in the attribute vector. We can choose M and the set of anchor boxes based on a statistical analysis of commonly occurring bounding boxes in images.

In any object detection algorithm, multiple overlapping boxes are output for the same class over the whole image. A post-processing stage of NMS is applied which prunes out boxes which overlap with a box with the highest score. Finally, note that the attribute vector is a mix of categorical variables and numerical variables. The loss function is a mix of the cross-entropy and the L^2 losses, and training is performed on a large dataset such as MS-COCO. The final result is a fast object detector suitable for real-time applications.

9.4 Attention in Computer Vision

In general, the encoder-decoder architecture with the attention mechanism, described in Sect. 3.2.4, fits well with computer vision tasks of image captioning, activity recognition, and object detection. In image captioning, the encoder CNN builds a representation R of an image which is used by a decoder RNN to produce a sequence of words describing the image. For activity recognition in a video, for each video frame, the encoder CNN creates a representation R, and the decoder RNN outputs the locations and class labels of activities. For object detection, an RNN-based decoder outputs a sequence of detections which have spatial correlation, such as in the case of people detection in crowded scenes.

For the three cases above, given an input-output pair (x, y), the architecture is given by

$$R = E(x),$$

$$y = D(R),$$

where E is the encoder CNN, D is the decoder RNN, and R is the feature blob output by the CNN. The attention is over different spatial positions of the feature blob. Attention to different temporal positions in a temporal sequence is qualitatively the same as attention to different spatial positions in an image. See the discussion in Sect. 3.2.4. See [191, 205, 247] to see how the encoder-decoder architecture has been employed in computer vision.

We discussed self-attention in Sect. 3.3. Self-attention is the basis for transformers used for NLU tasks; see Sect. 10.5. For NLU, the idea is to input word embeddings in a sequence to an encoder with several layers of self-attention in order to build hierarchical representations of the words. Using all-pairs attention, a global context is built, and therefore the higher-level layers build contextual representations of the words. Using self-supervised tasks, such as masked word prediction, a transformer is trained.

In object detection, CNN-based image features at different spatial positions of the feature blobs are analogous to word embedding vectors. An image is like a 2-D text document with visual words corresponding to different image patches, which can be masked during the self-supervised training. Feeding them into a transformer encoder-decoder allows a global context to be built based on relationships between different object instances. The encoder builds contextual representations of different parts of images. Therefore, if there is a bird which is occluded by a green blob, then it is more likely that the blob is foliage. The decoder outputs a set of predicted boxes based on the contextual representations which are output by the encoder. For details of this promising approach, see [25].

9.5   Semantic Segmentation

Semantic segmentation, which seeks to detect all pixels in objects of interest, is formulated by defining the ground truth for object detection at pixel-level resolution rather than at a coarse grid level. For each pixel, define the following attributes (assume that a pixel can belong to only one object):

1. Negative Class Label z. Same as before.
2. Positive Class Label Vector L. Same as before.

That is, we have discarded the bounding box attributes. A *semantic segmentation* CNN converts the $N \times M \times 3$ input into a segmentation mask of dimensions:

$$N \times M \times (C + 1).$$

This is a pixel-level classification task. It can be also viewed as an image translation task. Since it is a classification task, the cross-entropy loss is commonly used for semantic segmentation. One computes the cross-entropy loss for each pixel and then takes the average over all pixels. More customized loss functions have been also designed; an example is a loss function based on the Dice coefficient (which is related to IOU or the Jaccard coefficient) and which may led to better results when there is an imbalance between the size of the foreground object to be segmented and that of the background. The Dice coefficient is given as

$$D = \frac{2|X \cap Y|}{|X| + |Y|},$$

where X and Y are the ground-truth and predicted segmentation maps, respectively. In order to turn it into a loss function, Y will be the predicted real-valued probability map rather than the predicted binarized map, with multiplication and addition operations replacing the respective set intersection and set cardinality operations, and the coefficient will be subtracted from 1.

Several architectures for the pixel-level classification approach exist which are suitable for different computation budgets. One can adopt a fully convolutional approach in which all feature blobs computed by the hidden layers are of the same spatial dimensions as the image. That is a computationally expensive solution.

For practical computational budgets, the solutions are cast in the framework of encoder-decoder architectures because the intermediate layers have smaller spatial dimensions than the input and the output. The layer with the smallest size is called the bottleneck layer. There is downsampling followed by upsampling. Downsampling is due to the standard use of max pooling and convolution stride. Upsampling needs further work. Fortunately, downsampling followed by upsampling has been widely studied in the case of multi-rate systems and wavelet filter banks in the field of signal processing for a long time, where ingenious work has been done to design filters which are able to reconstruct the original signal. Conventionally in signal processing for a perfect reconstruction filter bank such as Haar or Daubechies wavelets, in order to upsample a signal which has been downsampled with an orthogonal filter bank, one inserts holes in the downsampled signal and applies the transpose of the downsampling filter. For orthogonal matrices, the transpose is the inverse. In CNNs, the classical wavelet upsampling method is known as the atrous (convolution with holes) and transpose convolution. The difference is that, for CNNs, the filters are learned. Though upsampling can also be performed by fixed filters such as bilinear interpolation, it is preferable to learn the upsampling filters.

An example of a semantic segmentation network is U-Net, which specializes in segmenting medical images and consists of convolutional layers with skip connections; see [180]. Skip connections improve the flow of information between different scales, which otherwise gets restricted by the bottleneck layers. An alternative way to view U-Net is as a multi-resolution architecture which is processing the image at different scales, and the resulting feature blobs at different scales get fused with a concatenation operation after going through up-convolution in a pyramidal fashion. Though an alternative way to fuse features is with an addition operation, it is preferable to use the concatenation operation and let the network learn how to process these features. Replacing 2-D convolutions with volumetric convolutions can be employed for segmentation of volumetric data such as medical images; see [143] for V-Net and see [38] for 3-D U-Net. The V-Net paper employs the dice loss. To handle foreground-background imbalance problem, 3-D U-Net employs the weighted cross-entropy loss with smaller weights for the background voxels. For these specific networks and in general for semantic segmentation, shape-aware loss functions can be designed which encourage matching of overall contour shape. V-Net uses residual blocks; 3-D U-Net can be also generalized to residual blocks. For other examples of AI models along with their sample code for medical imaging, visit [235]. Note that AI systems for medical imaging need to have user interfaces which enable physicians to correct the automatically generated contour by providing interactive guidance to the AI model. The AI model will use the guidance to automatically improve its contour in real time.

To get high accuracy results in medical imaging, segmentation alone is not enough because one has to first correct for motion artifacts if the organ or the

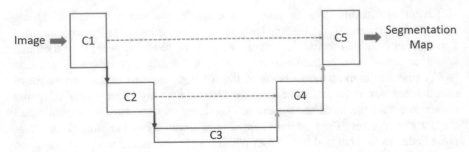

Fig. 9.4 A typical CNN model, shaped as U (or V), for semantic segmentation. The model can be viewed as operating at multiple resolutions of the image, from fine to coarse. The results are fused with concatenation. The downsampling and upsampling operations are best done by learned filters. Each block consists of several convolutional layers, which may be ResNet-like residual blocks. Another way to view the architecture is as a standard CNN with skip connections between different layers to improve the information flow between different resolutions. The image can be 2-D or volumetric (3-D)

patient moves, which is especially important in cardiac imaging. The detection and correction of these artifacts can be done by AI models. Interestingly, one needs to incorporate an MRI-imaging-specific layer in the architecture for data consistency in k-space because MRI data is captured in the Fourier domain by the scanning machine. This is an example of how novel research in customized domains extend the basic AI models to solve impactful real-world problems. The corrected image data is then input to a segmentation network such as U-Net. See [153] for the latest work on medical images using AI.

Fully convolutional networks (FCNs) [193] and SegNet [7] are two other examples of semantic segmentation networks. The underlying philosophy of down-sampling followed by upsampling, preferably with skip connections, forms the basis of AI models for segmentation; see Fig. 9.4. Skip connections facilitate the flow of fine-grained information from the downsampling stage to the upsampling stage. We also mention Tiramisu architecture [104] which uses densely connected blocks in a similarly designed architecture.

Suppose there are many cats in a given picture. We want to segment out the individual instances. Instance-aware segmentation is performed by detecting each instance using an object detection algorithm, and performing segmentation on each detected bounding box. Note that accurate segmentation will improve object detection, which in turn improves segmentation. They are two interlocked steps of the same process with the final goal of semantic scene analysis. Combining instance-aware segmentation for the foreground objects and semantic segmentation for the background is known as panoptic segmentation, which gives a rich, complete, and holistic scene segmentation.

> ### *** Exercise 25   Crowd Counting
>
> Consider the application of AI for counting the number of people in a large crowd at a sports stadium. The head or forehead of each person in the crowd is marked with a single dot in the training set. What will be a good way to formulate this problem? What will be a suitable loss function? See [220].

> *Object detection and semantic segmentation can be solved by using CNN as a blob converter which takes an input image blob and transforms it into a suitably defined attribute blob.*

9.6   Image Similarity

Fake news is a big problem in today's world dominated by the Internet. Let's say we detected a piece of fake news which consists of an image. To find the occurrences of the fake news in the online media, we will have to find near-exact duplicates of the image. It is an image similarity problem. Note that classical feature-based image-matching solutions can also be developed to solve the problem.

An image classification CNN works as a feature extractor because the activation patterns of layers can be viewed as the features of the input image. To see whether two images are similar, we just measure the similarity of these features (activation patterns). The cosine distance is a popular and robust way to measure high-dimensional vectors and can be employed as the similarity metric.

Suppose you have a database of images and your goal is to see whether a given query image is similar to any of the database images. Suppose the database has one billion images. How will you implement approximate nearest neighbor search? One technique which has been used by Flickr is *locally optimized product quantization*; see [17, 107], in which a high-dimensional vector is split into sub-vectors. Each half can be clustered separately using the k-means algorithm. Therefore, for a new feature vector, we compare it with $2k$ clusters of the database feature vectors. If we had done clustering on the original dimensions, we would have needed k^2 clusters for the same cluster granularity, i.e., for the same number of images in each cluster.

Explicitly training neural networks to excel in pairwise comparisons can be employed where there is enough training data. A Siamese pair of CNNs with weight sharing is used, which takes a pair of images as the input. The outputs of the two parallel networks are compared with a distance metric or are concatenated and then fed into additional learnable layers. The whole network is trained E2E on positively (similar) paired images and negatively (not similar) paired images. Therefore, the architecture looks like the following:

$$R_1 = F(I_1),$$

$$R_2 = F(I_2),$$

$$d = \text{dist}(R_1, R_2),$$

where I_1 and I_2 are the two images and R_1 and R_2 are the representations of the two built by the CNN. For database-matching problems, instead of having the same CNN, one can use two CNNs F_1 and F_2 for query image I_1 and database image I_2, respectively. The Siamese network may be applied in the general self-supervised learning to learn latent representations; however, since too many dissimilar pairs are required to build a negative dataset, their use in contrastive self-supervised learning is practically limited.

A triplet combines a positive pair and a negative pair with a common anchor:

$$(A, B, C),$$

where (A, B) is the positive pair and (A, C) is the negative pair with A as the anchor image. A triplet is input to three copies of the same CNN, which compute features of A, B, and C, respectively. The loss function ensures that

$$d(F(A), F(B)) \leq d(F(A), F(C)) + \beta,$$

where d is a distance metric such as the Euclidean distance, F is the feature mapping defined by the CNN, and $\beta > 0$ is a margin hyperparameter which is used to discourage the network from outputting zero all the time. Triplets should be chosen carefully so that they are difficult examples. See Fig. 9.5. The Siamese network trained with the triplet loss provides a solution to face recognition in which a query image is matched against a database of images. One extra detail for face recognition is that the face images are typically preprocessed (aligned and normalized) for pose invariance:

$$d(F(P(A)), F(P(B))),$$

where P is a preprocessing module, which can be implemented by an AI model using an autoencoder or by using a separate component. In fact, it is possible to build an end-to-end solution. Besides many-to-one processing which maps a face image to a canonical frontal view, one can generate multiple poses of faces which will constitute one-to-many processing. See [223] for further reading on face recognition. It is very interesting to observe that one can use AI to recognize faces of brown bears. See [39] for such a work on face recognition in which the match between the embeddings of bear faces computed by a CNN model is performed by an SVM.

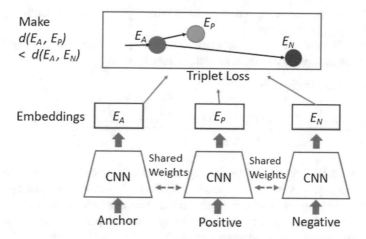

Fig. 9.5 Image similarity using triplet loss. The distance between the embedding of the anchor image and that of the positive image is made much smaller than the distance to that of the negative image. This provides a solution to face recognition. During training, three copies of the CNN model are used with tied weights, and that's why it is called the Siamese architecture. The extra work which is needed for face recognition is preprocessing of face images. This processing typically includes cropping, normalization, and alignment to a canonical frontal view

9.7 Video Analysis

It is fair to state that the most popular form of visual content is videos. YouTube, Netflix, Amazon Video, and Disney Channel are all examples of large enterprises to serve our constant need to consume videos.

In this section, we develop intuition how one can use AI for videos. Let us start with the observation that a video is a 3-D image. Taking k RGB frames as $3k$ channels, we can work with videos. One can generalize the 3-D convolution for images to *pure 3-D* convolution. For images, this is not pure 3-D as the convolution filter is of size $w \times h \times C$, where C is the number of feature maps; see Fig. 3.2 and Sect. 3.1.3. In pure 3-D convolution, the filter is of size $w \times h \times d$ where $d < C$.

This seems fine except there is the problem of the scarcity of data. Since k may be something like 500 for a 20-second video clip, the raw pixel space has k times as many dimensions compared with that for still images. Although, in theory, manifolds exist waiting to be carved out in such a space, practically speaking there are not enough real-world videos.

Therefore, k is reduced to a small number, and short segments of videos are analyzed, each of which lasts for a couple of hundreds of milliseconds. Over a video scene lasting tens of seconds, a sequence of results is obtained which has to be pooled, fused, and post-processed. A natural idea is to combine it with sequential AI models such as RNNs. The system is broken into two components:

1. A CNN-based computer vision module which processes consecutive, possibly overlapping, video segments
2. An RNN-based temporal sequence processing module which combines the results of the computer vision module across time.

The above can be written as

$$(z_1, \ldots, z_t) = F(x_1, \ldots, x_t),$$

$$y = G(z_1, \ldots, z_t),$$

where the CNN F works independently on t video segments x_i's and the output sequence gets processed by the pooling RNN G. There are pooling approaches which are simpler for practical applications. For example, if you are trying to detect whether there was a particular activity, for example, of a dog running in a playground, global max pooling will work fine.

Can we do better? The idea of helping AI with human intuition, prior knowledge, and inductive bias can result in significantly better results with less data. We know a priori that motion is an important feature of a video and a motion field tells us a lot about what is happening in the video. Instead of making an AI model learn how to compute this motion field, it is precomputed. Therefore we have two types of data:

1. The raw video frames.
2. The precomputed dense motion field such as optical flow.

The CNN will receive the two input streams and process them in parallel. At some stage, it will merge the two computation paths; this can occur right at the input layer, or at the end just before the final output or somewhere in the middle. This can be expressed as follows:

$$m = \text{OpticalFlow}(x),$$

$$y = F(G(x), H(m)),$$

where G and H are two processing streams for the video data x and the motion field m, the outputs of which get fused at some stage and processed together by F. The fusion can be implemented as concatenation $G(x) \oplus H(m)$ or addition of feature blobs $G(x) + H(m)$, the best method determined by experiments for the particular problem.

Video analysis is an example in which end-to-end learning is not always the best idea, and it is preferable to break up the system into components. The idea of augmenting the input, here with the optical flow, is referred to as *enriching* the input. See [199] for an example of how videos can be analyzed using optical flow.

In a number of applications, objects of interest need to be tracked and their poses estimated in videos. Object tracking can be efficiently performed by classical computer vision algorithms, where corners or keypoints are tracked. For recent work which uses AI, see [221] for 3-D pose tracking of common objects in RGB-D videos for robotic applications. The tracking is based on keypoints, which are learned by employing AI models to compute image features and to generate keypoints based on these features. For an application which is of great value for behavioral studies of wildlife, see [185] for tracking of animals, where Mask R-CNN, a region-based CNN model for object detection and semantic segmentation, is applied.

9.8   3-D Data

There are two types of 3-D data: (1) volumetric and (2) point cloud.

The first type is volumetric and consists of voxels. This is truly a 3-D image and CNN readily generalizes to it; see also the discussion in Sect. 9.7 where we considered pure 3-D convolutions. Consider the case of CT images in medicine. Let's say we want to detect regions of interest in a pulmonary CT scan. One approach is applying 3-D CNN. An observation is that we want AI to learn invariance to certain geometric transformations such as rotations and reflections. Therefore, instead of applying one copy of a convolution filter on the input feature map, one applies rotated and reflected copies of the filter; see [11], as in group convolutions (dividing the input into groups and applying filters to them separately). This provides an alternative to the data augmentation approach performed by rotations and reflections of the input data. For an application of this scheme to CT-scan volumetric data, see [243]. For segmentation of volumetric medical images, a multiscale architecture works well in which volumetric 3-D convolutions are performed at different resolutions. For example, see [143], where a loss function based on the Dice coefficient, parametric ReLU, and data augmentation is successfully applied to a multi-resolution CNN for MRI data.

The second type of 3-D data is a 3-D point cloud. Self-driving cars and robotics are technologies which will become commonplace one day. It is possible that just by using RGB color cameras without any depth sensors which capture 3-D point cloud data, they will be able to move around. Still, having a depth sensor is helpful irrespective of the future computer vision breakthroughs in the RGB space.

An approach to process point cloud data is to convert it into volumetric data. This voxelization is a workable solution, but there is a loss of information if quantization is coarse. By increasing the 3-D voxel grid size, the problem can be solved at the cost of computation. Another approach is based on a key observation that occupancy of the grid cells is low and it is very sparse data. Therefore the computation is significantly speeded up by applying a convolution filter only on the active grid cells; see [64]. Therefore, we have $y = F(V(x))$, where the 3-D data x is voxelized by the operation V, and F is a 3-D CNN which applies a convolution filter only on the grid cells which are occupied.

Another solution is to project the 3-D data to 2-D [169] from multiple views and use the standard 2-D CNN on the projected images. The projection is analogous to the Radon transform, which is used to construct 2-D images from 3-D medical

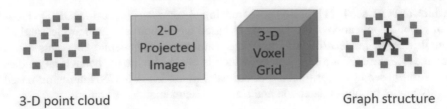

Fig. 9.6 Three different approaches to represent and process 3-D point cloud data shown at the left end: (1) 2-D projected image, (2) 3-D volumetric data in an occupancy grid, and (3) 3-D graph with local neighborhood concept shown in red for a point. For the 2-D image form, a 2-D CNN can be used. For the 3-D grid method, if many voxels (occupancy cells) are empty, then sparse pure 3-D convolutions are used. For directly working with 3-D points in the form of a graph, a graph CNN is employed

imaging data. Sometimes it is easy to build a 2-D image from a point cloud dataset. Consider the example of self-driving cars. Place yourself in the middle of the point cloud from where the data was captured using a rotating vertical stack of lidars. Each lidar sweeps the scene horizontally giving a row of the 2-D image. There are as many rows as the number of lidars. Using this spherical projection, we get a 2-D range image. Therefore, we have $y = F(P(x))$, where the 3-D data x is converted into a 2-D image by the projection operation P, and F is a 2-D CNN.

Finally, an alternative solution is to work with points directly and make use of techniques similar to graph convolutional networks; see Sect. 3.5 and [167]. The points are vertices of a graph with edges to nearby points. There is a concept of locality as determined by the distance metric, and therefore graph CNN can be used to extract local features in the neighborhoods which are grouped into larger neighborhoods to extract higher-level features. For each point x, we have a neighborhood convolution $F(N(x))$ where $N(x)$ is the local region around x. This is similar to sparse convolution on voxelized data, where in the latter the convolution is over a quantized, uniform, grid-like neighborhood. See Fig. 9.6, which shows different ways in which 3-D point clouds are processed.

An earlier version of the point-based approach known as PointNet [168] does not capture the local structure and instead builds learned independent N-dimensional representations of the points using a feed-forward neural network followed by a global max pooling to yield the global representation, which is input to a classification network. To make PointNet robust to different poses of the object, *spatial transformer network* layers are employed while building the representations. A spatial transformer network can be viewed as an image translator which performs regression on the latent parameters of the pose, such as on the latent affine or homography transformation coefficients, and then creates a transformed image corrected for the pose. It can be added as a layer to any classification network to make it more robust to changes in poses; see [102].

> *AI techniques for still images can be generalized to videos, 3-D volumetric data, and 3-D point cloud data by making use of the specific nature of the data.*

9.9 Self-Driving Vehicles

A real-world use case of AI for computer vision is the technology for autonomous vehicles. Let's get high-level intuition into how AI is employed to build this technology based on the tools which have been introduced in this book.

A self-driving car is equipped with the following at different places in the vehicle:

1. C color cameras.
2. L lidars.
3. R radars.
4. S ultrasonic sonars.

Where to place these sensors and what should be the values of these hyperparameters? Experiments are underway. Can we have $L = 0$? Yes, and that's a topic of debate and the future will provide a definitive answer. Humans can infer depth from stereo cues such as disparity map and monocular cues such as texture and lighting. Depth is an important derived feature which assists humans to quickly segment out the scene into objects. Therefore, irrespective of whether $L = 0$ or not and of whether it is directly measured or algorithmically computed, depth will be used extensively in the future computer vision applications including self-driving cars. If $L = 0$, one can employ AI models to compute depth from the camera images. To obtain a large training data, it is possible to use self-supervised learning without any need for manual annotation because the computed depth has to be consistent with all the camera views. The inconsistency can serve as a loss function to be minimized.

Let's assume we have all the sensors. The cameras provide conventional video streams. They offer excellent details of the scene but don't work well in adverse weather conditions and at nighttime. They don't provide depth information, which is needed to avoid obstacles. Stereo vision can provide depth information, but for that you need compute intensive algorithms working on textured images. Lidars, most expensive of all, are synchronized with the cameras, and they use infrared light pulses which sweep the surroundings giving us a 360-degree point cloud. They can detect nearby small objects and work in all lighting conditions. Though their resolution is less than that of cameras, lidars can measure velocity and give you an idea of the general shape of objects. Note that the point cloud will become sparse for faraway objects. The lidars, which sit on the rooftop of the car, constitute a stack of lidars, each of which does a planar sweep (which may be tilted at an angle). The intensity of the reflected light pulse is also measured, which provides additional information about the surface of the object. Lidars can get affected by bad weather, and they can detect small particles in the air, leading to false alarms. Radars are reliable and impervious to bad weather conditions, they provide a very coarse low-resolution idea of objects due to the high wavelength of radio waves, and they can measure the speed of objects. Sonars use sound waves and are typically used when you are parking your car.

Note that sensors can malfunction, and environmental conditions, for example, water or dirt on the camera, can degrade the quality of data. They have to be calibrated carefully so that the mapping from the raw data signals to the processed data with spatial coordinates is accurate.

Suppose there are 64 lidars on the rooftop and the resolution is 0.25 degrees. That means that in a full 360-degree sweep, a lidar will provide 1,440 points. Using the spherical projection method discussed in the previous section, we will have a $64 \times 1,440$ projected range image. Scan lines of lidars become rows of the range image. Note that some pixels in the range image may have missing depth values. We can also build a bird's-eye view of the scene by looking down at it from a height, and dividing the scene into small cells and collecting statistics of points inside each cell, such as density, mean intensity and maximum height.

As discussed in the previous section, there are several different ways to work with 3-D lidar data: (1) as a 2-D range image, (2) directly as 3-D points, and (3) as volumetric data. See Fig. 9.6, which shows these three representations. The processing of lidar data has to be fused with that of RGB data at some stage, which could be early, middle, or late fusion. That is,

$$y = F(G(x_{\text{Camera}}), H(x_{\text{Lidar}})),$$

where G is the AI model for RGB data, H is for Lidar data, and F is the fusion model. The fusion could be implemented as concatenation or addition, depending on the architecture, and as determined by experiments. Similarly, radar data should be processed and its results fused to make final decisions. GPS/IMU sensor data is also processed and integrated with signals from other perception components.

The goal of perception in autonomous vehicles is holistic scene understanding, with particular emphasis on detection of pedestrians, vehicles, lane markings, drivable area, alternate drivable area, traffic lights, traffic signals, traffic intersections, and obstacles. See Fig. 5.6 for application of multitask learning to perform these multiple detection tasks. In addition to their location, velocity and acceleration of objects have to be computed. Therefore one can employ AI models for object detection, panoptic segmentation, and image classification to solve these problems. As an example, to read a traffic light, we have

$$x_{\text{Light}} = F(x_{\text{Camera}}),$$

$$y = G(x_{\text{Light}}),$$

where F detects the location of the traffic light and returns the image region x_{Light} containing the light, and G classifies this region into one of the three categories:

$$y \in \{Red, Green, Yellow\}.$$

Object tracking and motion analysis, combined with the velocity measurements from lidars and radars, are employed to provide a full picture of the dynamic scene. High-definition maps can be taken as priors in perception.

Therefore we can appreciate the challenge which the technology has to overcome. At the highest level, the perception component has to build an accurate world model W_t for the current time t, in diverse weather and lighting conditions, which has location as well as motion information of vehicles, people, and pets in the surroundings, and which has information about the traffic signs and signals:

$$W_t = F(x_{Camera}, x_{Lidar}, x_{Radar}, x_{GPS}, x_{IMU}, x_{Map}, W_{t-1}),$$

based on all the incoming data and all the priors which are available. Based on the output world model W_t, behavior of other vehicles and pedestrians has to be predicted, the ego vehicle has to be controlled, and the path to the destination has to be dynamically updated. The quest for the best possible F is on. Clearly, F has to be divided into different components which all have to work together, and the whole system has to be debugged and fine-tuned carefully. One natural way to divide F is by using *multitask* CNN-based AI models for the perception task whose output becomes input to a higher-level temporal recurrent module. Note that there are multiple interdependent perception models for multiple streams of sensor data. Finally, an AI model can be trained on the outputs of the above modules to produce a bird's-eye view of the scene which can be displayed on the dashboard. The bird's-eye view can be considered as a human-interpretable, graphical, and compact representation of W_t.

Perception is a key part of self-driving cars, and AI models which work with images and videos have to be extended to 3-D data. See the surveys [5, 72] to learn how 3-D data is being used to solve perception problems for autonomous vehicles.

The results of scene understanding become the input to the subsequent components which implement behavior prediction (B), path planning (P), and vehicle control (C),

$$(B_t, P_t, C_t) = G(W_t).$$

Prediction of the behavior of different agents in a multi-agent dynamic setting within the context of the traffic conditions can be formulated as a supervised learning problem. This can be achieved by looking at trajectories of pedestrians and other vehicles in a bird's-eye view image. Alternatively, one can work with a vectorized representation of these trajectories and use a hierarchical graph neural network to model spatially local interactions of different agents; see [54]. Techniques from reinforcement learning and imitation learning are applicable to path planning because the car has to perform sequential decision-making. The decisions have to be mapped to commands to the car to control its motion. Vehicle control is defined as control of the motion of the vehicle, for which supervised learning and reinforcement learning can be employed. See survey [118] to learn how deep learning can be used for vehicle control. To learn more about the autonomous driving application of deep learning, see [8, 70, 249].

Self-driving cars need huge amounts of training data, which should include difficult and rare situations. Therefore proper management of data is crucial. The real-world data can be provided by the cars being driven. By building the capacity to acquire data with the desired attributes, for example, of cars with mounted bicycles or of occluded road signs, using an automated or a semiautomated method, the technology can improve while it is being used. This is an example of *active learning*, in which one intentionally seeks out difficult examples to be included in the training set.

> *Self-driving cars depend heavily on AI to solve perception problems. Since they depend on 3-D sensors in addition to color cameras, there has been significant interest in making AI-based computer vision work on 3-D data. Multiple solutions exist to build self-driving cars and the optimal configuration has yet to be figured out.*

9.10   Present and Future

There is a significant amount of published work on the application of AI to computer vision which represents the state of the art. Research papers on image classification often use CIFAR-10, CIFAR-100, and ImageNet datasets [47, 117]. For object detection and semantic segmentation, MS-COCO dataset is popular [133].

Here are some references for further reading which have inspired the contents of this chapter. See the ImageNet website [232] for details about winners of the competition. For examples of innovative work in CNN, see [79, 116, 124, 211].

In object detection, see [190] for sliding window-based approach, [178] for region proposal-based approach, and [135, 176] for a single-shot regression-based approach. A common variation of AI models for computer vision has been to work with the input or the features at different scales (resolutions). See [131] for a multi-scale approach known as *feature pyramid network* for object detection. The feature pyramid network uses the philosophy of semantic segmentation networks which have a downsampling stage, an upsampling stage, and skip connections. The object detection is done at each upsampled feature map. Skip connections allow the information to flow from high-resolution, semantically weaker, and lower-level layers to low-resolution, semantically stronger, and higher-level layers. Since it is easier to obtain background negatives due to the very large number of negative samples compared with the foreground positive samples, by correcting this imbalance with a modification of the loss function on a sparse set of difficult examples, one can improve the working of single-shot object detectors; see [132] for RetinaNet.

For semantic segmentation, see [180] for U-Net, [7] for SegNet, [193] for FCN, and [29] for DeepLab. For the instance-aware case, see [130]. In [78], faster R-CNN is extended to output a segmentation mask for each detected bounding box, and the approach is called Mask R-CNN. It provides a solution for instance-aware segmentation at a coarse grid level.

For image similarity and its application to face recognition, see [94, 187].

Note that different computer vision tasks, namely, image classification, object detection, semantic segmentation, instance-aware segmentation, panoptic segmentation, and image similarity, overlap with each other and they all are interrelated offshoots of the same task of *holistic scene understanding*.

Before 2012, in ImageNet competitions, top-1 and top-5 accuracy numbers of classical computer vision techniques were in the low 50s and low 70s (in percentage), respectively. In 2020, the percentage numbers have increased to the high 80s and high 90s, respectively.

Therefore, it is natural to look beyond the existing techniques and ask how one can build next-generation AI models which overcome the shortcomings of the current approaches, such as adversarial examples and dependence on large datasets. Though CNN-based models have translation invariance, they lack general geometric invariance. It is anticipated that by incorporating techniques from the way human vision works and by making use of ever increasing computing power and data, we are bound to enter a new era in computer vision.

9.11 Protein Folding

Finally, it is worth mentioning that techniques in this chapter apply to any problem in which one can obtain a visual representation. Consider the problem of prediction of 3-D structure of protein sequences. Suppose a given protein sequence S is made of L amino acid residues:

$$S = a_1 \ldots a_L.$$

The goal is to predict how S folds into its tertiary structure. We can build an $L \times L$ "input image" X of covariation of the residues. $X(i, j)$ is the covariation of a_i and a_j. A residue a_i is supposed to covary with residue a_j if it has been experimentally observed that, in many similar protein sequences, the two residues mutate and change together. The "output image" Y is the predicted $L \times L$ pairwise distance map between the residues. It can be a k-channel image if the prediction is a k-bin distance histogram. $Y(i, j)$ is the k-bin distance histogram for a_i and a_j. The input image X can be also made a multichannel image by adding other features. A one-dimensional $L \times 1$ sequence feature can be turned into a 2-D channel by tiling it repeatedly across columns. Then, a CNN F can be trained to implement $Y = F(X)$. The output Y can be viewed as a smooth constraining function on the geometry of S, and gradient descent on this function can be used to predict the 3-D structure of S. This idea serves as the basis of AlphaFold; see [189]. In AlphaFold, torsion angles are also predicted besides distances. By replacing CNN with more advanced AI models, further progress in solving this important problem in life sciences and medicine is certain. Instead of using the direct method of X-ray crystallography which is expensive and slow, we can then employ AI to quickly obtain 3-D structure of proteins.

9.12   Chapter Summary

Computer vision is a widely studied application of AI, and the last few years have seen explosive growth in techniques to solve problems of image classification, object detection, image similarity, and semantic segmentation. For image classification, the ImageNet competition provided a strong incentive to build innovative CNN architectures. A pre-trained model can solve a variety of practical problems with minimal effort. Object detection can be solved by reducing it to image classification. A regression-based, single-shot approach formulates object detection as a regression problem, leading to fast real-time object detectors used in practice. Image similarity can be solved by using CNNs as feature extractors and by training Siamese networks under the triplet loss. Semantic segmentation is pixel-level object classification in which the output is a semantic map of the same resolution as the input. Besides still images, techniques have been developed for videos, 3-D volumetric data, and 3-D point cloud data. For videos, providing motion field, in the form of an optical flow, as additional input helps the AI model learn with less data. For 3-D volumetric data, pure 3-D convolution can be applied. For 3-D point cloud data, a variety of techniques have been experimented with. Effective techniques for point cloud data make use of the fact the data is sparse in the surrounding volume. All of the above techniques are heavily used in the design of self-driving cars, which have to process multiple streams of data to produce a holistic world model of the surroundings of the ego-car in order to predict the behavior of other cars and of people, to control the vehicle, and to plan the journey ahead. The challenge for autonomous vehicles is to make it all work in different lighting and weather conditions and to be able to handle rare situations.

Chapter 10
Read, Read, Read

> For thought is a bird of space, that in a cage of words may indeed unfold its wings but cannot fly.
>
> Kahlil Gibran

> ... it is the words that sing, that soar and descend, I bow to them, I love them, I cling to them, I pursue them, I bite them, I melt them down, I love words so much ...
>
> Pablo Neruda

There is one thing which humans do which no other animal does.

Animals are smart with quick reflexes and special abilities. They learn from others and from their own experience, they use tools, and they can navigate complex social hierarchies. It is only humans who save their accumulated experiences and observations for eternity. The shared store of knowledge is what enables us to accomplish great things. This cultural wealth primarily exists in written form which passes from generation to generation and grows over time. Without text, human civilization would regress back by 5,000 years.

The ability to read and to write is an integral component of our intelligence. Therefore, AI can't be called truly intelligent if it can't read or write. Recognizing dogs and cats and obeying the spoken command "Play old Bollywood songs" are fine, but it is in the magnificent power of language we will see glimpses of intelligence. Being able to converse with us will be a core component of the Turing test for robots in the future.

In this chapter, we will develop intuition about AI-based NLU.

10.1   Natural Language Understanding

Consider the following NLU tasks:

1. Is this product review positive, neutral, or negative?
2. What is the category of this document: legal, financial, scientific, or other?
3. Given a collection of documents, is a new document similar to a document in the collection?
4. Given a collection of history books, answer this question: "Why did World War I start?"
5. On an Apple technical support website, you initiate a conversation with a chatbot by asking "How do I capture a part of my screen on my MacBook?"
6. Can you summarize this long legal document?
7. Can you translate this English sentence into German?
8. Given recent posts on the social media, has the approval of the government gone up or down?
9. Automatically grade this essay written by a Grade-8 student.
10. Find those news stories in a given news corpus which talk about the positive economic impact of immigration that are backed by scientific studies and surveys.
11. Detect the biomedical entities such as "62-kDa protein (p62)," "serotonin syndrome," and "hepatitis B surface antigen" in a medical research article, and find their types. (Their respective types are gene, disease, and chemical.)
12. You write a book on AI. After finishing the book, you ask an AI model to suggest a title, a subtitle, and a summary of the book.

It is not obvious immediately how to solve the above problems. Certain tasks seem more tractable, and by using counting words and phrases and by having dictionaries and thesauruses, we can provide a solution. For example, we can collect three sets of words, positive, negative, and neutral, for the sentiment classification problem. At the other end, it is not clear how to create an informative, succinct summary of a legal document and how to implement a chatbot. We can use the following classical Natural Language Processing (NLP) tools to get started on the problems.

- Pre-processing of words to prune out the common words and to create the base forms of words.
- Extracting n-grams from a document. An n-gram is a short sequence of n co-occurring words in text.

- The term-frequency and inverse document-frequency (tf-idf) method to find informative words in each document.
- The named-entity recognition (NER) method which detects words describing people, places, organizations, date, time, money, and other named entities.
- The part-of-speech tagging method to label words as nouns, adjectives, verbs, and other grammatical constructs.
- Parsing of sentences according to the grammar.
- The latent semantic analysis method on a word-document matrix to build a lower-dimensional space of latent concepts for document clustering.
- The topic modeling method by using clustering techniques or by the statistical generative technique known as latent Dirichlet allocation in which a topic is a distribution of words and a document is a mixture of topics.

The goal of this chapter is to develop insight into AI-based tools which have not only enhanced the existing tool set but have led to effective solutions of challenging problems.

10.2 Embedding Words in a Semantic Space

The starting point for NLU solutions is the representation learning of words. Words are the units of a language, and complex concepts and thoughts are hierarchically built on top of them. We *embed* words in a high-dimensional numerical semantic space commonly known as the embedding space. A word v becomes a D-dimensional vector $E(v)$ in the embedding space, where D is typically in the hundreds. This is called a dense representation,

$$[v_1, v_2, \ldots, v_D],$$

when contrasted with a sparse V-dimensional one-hot representation,

$$[0, 0, 0, \ldots, 0, 1, 0, \ldots, 0, 0, 0],$$

in which there is exactly one 1 at the numerical index of the word in a vocabulary having V words. Typically, V is in the hundreds of thousands. See Fig. 10.1. Embedding words in a semantic space is an example of a distributed representation. Representing a symbolic entity such as a word or a concept by the activation pattern of several neurons is called a distributed representation. A neuron can contribute to the representation of many entities, and therefore it is a many-to-many mapping.

D dimensions of the embedding space signify latent semantic concepts, which are automatically discovered. A word is a combination or an activation pattern of these latent concepts. The embeddings of words reflect relationships between words, and similar words will have similar activation patterns. Words don't exist in total isolation as discrete entities. A word works with other words to create concepts.

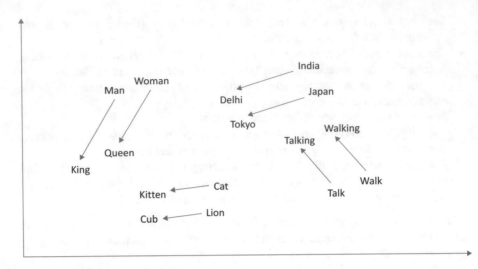

Fig. 10.1 Embedding words in a high-dimensional semantic space. Words which occur in close context in a text corpus have embedding vectors which are close to each other. Directions in the space have a semantic interpretation. This is an example of a distributed representation of concepts

Our thoughts and concepts reflect the world and our inner experiences, in which things are connected to each other. We say that words are semantically related to each other if they typically occur close to each other in a text corpus. They share the same context, and therefore they will share the activation pattern of underlying concepts. Word embeddings are computed by taking a large text corpus consisting of Wikipedia, a news corpus, and books and feeding it to a training algorithm. Two widely used embeddings are Glove and word2vec. For Glove, co-occurrence statistics of words are used to find embeddings. For word2vec, in one of its versions, called the *skip-gram* method, a window of adjacent words around each word w is used, and the goal is to find their embeddings such that their softmaxed dot products with the target word w are maximized. The conditional probability of a context word u given the word w is defined as

$$P(u|w) = \frac{\exp(E(u)^T.E(w))}{\sum_v \exp(E(v)^T.E(w))}.$$

The average probability is maximized over all such pairs. Starting with random embeddings, the vectors are updated iteratively to maximize this objective function. Therefore, there is an implicit contextual language modeling task which is being used to train the embeddings. In another version of word2vec, called continuous bag-of-words, the average of embeddings of context words is used to predict the target word.

The result is a $D \times V$ matrix E' in which embeddings are arranged column-wise. Multiplying the one-hot representation v of a word with E', that is, the matrix-vector product $E'v$, yanks out the corresponding column, and we get its dense embedding vector. Since an FFN layer is matrix multiplication, therefore an embedding layer is implemented as the very first hidden layer with V input neurons and D output neurons. The embedding layer converts the one-hot representation into the dense representation, which is then processed by the subsequent layers. The embedding layer is initialized with pre-trained Glove or word2vec embeddings and is trained by backpropagation to fine-tune it further. See [159] for Glove and [142] for word2vec.

In general, the idea of embeddings is applicable to any set of categorical variables to capture their semantic similarity. Therefore, an embedding is a learned function,

$$E : X \to \mathbb{R}^D,$$

where X is a set of categorical entities represented as one-hot vectors, and the mapping is such that the contextual semantic inter-relatedness between these entities, as measured by their co-occurrences in the same context, gets mapped to spatial relationships in D-dimensional Euclidean space.

The method of using embedding vectors to represent words for language modeling as a special case of distributed representation was introduced in [14] for language modeling tasks. Embedding is featurization of a word, and it is an internal representation built by a neural network. An AI model builds a hierarchy of representations which collectively form the distributed representation of its input. Therefore, if we look at representation learning in the most general manner, it is not only the first layer but every layer which is building embeddings.

> *A powerful idea in AI-based NLU is to build the distributed representations of words in a semantic space. A word is an activation pattern of underlying latent semantic concepts (neurons). These distributed representations can be compared to measure semantic similarity of words, and they provide the starting point for further NLU processing.*

10.3 Sequence to Sequence

Pre-trained word embeddings can be used to solve NLP problems. For example, consider classification of a sentence S having words $(w_1, \ldots, w_n)$. One can take embeddings of individual words and take the mean vector,

$$E(S) = \frac{1}{N} \sum_{i=1}^{N} E(w_i),$$

as an embedding of the sentence S in a naive approach. One can train an AI model to classify the mean vector. For advanced approaches, we have to train the whole model in an E2E manner. For sequential data, an encoder-decoder architecture is a general model which can be trained to solve NLU problems. See Fig. 3.5 and Sect. 3.2.4,

10.3.1   Encoder-Decoder Architecture

In Sequence-to-Sequence (Seq2Seq) problems, the part of an AI model which creates a representation of a chunk of the input sequence is called the encoder. The decoder takes the representation and produces the output sequence.

If the input is a chunk of spatial data, the encoder can be a CNN. If the data has a sequential structure with full temporal context, it can be a bidirectional RNN (with LSTM blocks). The line between sequential data and spatial data is not rigid because the spatial and temporal dimensions are similar. Since a sequential data with full context can be viewed as spatial data, therefore a CNN can be creatively used for the encoder as well. If full context is lacking, a CNN with causal convolution filters can be employed. For online applications, with no future context and with long-term dependency on the past, it is natural to choose the encoder to be a multi-layer LSTM.

Several design choices for the decoder exist depending on the problem. If the output is non-sequential, then the decoder can be tightly integrated as additional layers on the top of the encoder, with the distinction between the encoder and the decoder getting blurred, for example, in the case of an autoencoder. If the output is sequential and unbounded, the decoder is an RNN (with LSTM blocks), and the next output of the decoder depends not only on the encoder output but also on the decoder's previous output.

10.3.2   Neural Machine Translation

To illustrate the working of an encoder-decoder architecture, let us take the concrete example of Neural Machine Translation (NMT). Suppose you have to translate the following sentence in Hindi into English,

$$\texttt{tum kahaan ja rahe ho?}$$

We have five words and a question mark. We compute their D-dimensional embeddings,

$$v_1, v_2, v_3, v_4, v_5, v_6,$$

and feed them one by one into an RNN. The last hidden state of the RNN is the encoded representation R of the input. Suppose R is K-dimensional, and therefore a temporal sequence of six D-dimensional points in the embedding space has been mapped to a point in a K-dimensional "sentence" space,

$$R = E([v_1, v_2, v_3, v_4, v_5, v_6])$$

where E is the encoder. The encoder has created a distributed representation of the sentence in a semantic space. From word embeddings, we have gone to sentence embeddings. We can do better by using a bidirectional RNN because we have the full temporal context. Instead of using the last hidden state, the hidden states of all time steps are aggregated together to form a learned representation R.

The decoder D is given a start signal and it consumes R and outputs "where,"

$$\text{"where"} = D(R, [Begin]).$$

It has learned during training to output "where" for points in the neighborhood of R. The decoder actually outputs a classification probability vector of the same size as the vocabulary, and the maximum occurs at the word "where". At the second time step, the decoder uses R and its output "where" to produce "are",

$$\text{'are'} = D(R, \text{'where'}).$$

Note that we are supplying R at every time step. This is optional and can be omitted without affecting the performance of the algorithm. At the third step, the decoder uses R and its latest output "are" to produce "you." Note that at every time step, the decoder implicitly encodes the history of its previous steps in its hidden state. Finally, we obtain the full English translation,

Where are you going?

In the above, we chose the word with maximum probability as the output at each decoding stage. In *beam search* at inference time, the top k words are selected, each of which leads to the next k words; therefore a search tree is built from which the best overall translation is chosen that maximizes the product of probabilities across the sequence. To incorporate *beam search* at the training time, see [244]. Finally, note that the number of words in the original sentence and in the translated sentence may be different because the decoder works after the encoder.

10.4   Attention Mechanism

Let's continue the Hindi to English NMT example. Now consider the sentence,

Maine ek AI ki kitaab padi jis mein ek accha udaharan hai. Us
mein kai kathin prashan hein.

It translates to the following sentence,

> I read a book on AI which has a nice example. It has several
> difficult exercises.

When AI is producing the English word "which," it should pay attention to "book" and "AI," two candidates for the interrogative pronoun. When the word "it" is produced in the second sentence, it should pay more attention to "book." When producing "exercises," attention should be really on "book," and the translation "exercises" should be preferred over "questions."

This is achieved by the attention mechanism; see Sect. 3.2.4. Past hidden states of the encoder are saved in a memory bank called the attention memory, and the attention mechanism selectively retrieves the contents of the memory during decoding to produce better results. The intuition is to pay more attention to relevant words from the past. Therefore, for example,

$$S(h_{\text{it}}, h_{\text{book}}) > S(h_{\text{it}}, h_{\text{nice}}),$$

where S is the attention score. Let's say the memory bank has N hidden states, each being a K-dimensional vector. A simple method which works well in practice is to implement attention by taking the current hidden state of the decoder, computing its similarity score with each of the N vectors in the attention memory, and taking a weighted combination of N vectors. The score function can be implemented by a neural network and learned during training.

This section is inspired by the latest work on NLU; see [9, 137, 218, 245]. Furthermore, see [34, 206] to learn how encoder-decoder architectures have been employed in NLU.

10.5 Self-Attention

This is a generalization of the attention mechanism, which was introduced in Sect. 3.3. In the standard attention, we have a vector u and attention memory $\{w_1, w_2, \ldots, w_N\}$ such that

$$u \notin \{w_1, w_2, \ldots, w_N\}.$$

In self-attention, we set $u = w_j$ for $1 \leq j \leq N$ and perform pairwise attention; therefore it is called self-attention.

In transformers, which are based on self-attention, see Fig. 10.2; there are three intermediate vector representations of each word w learned from its embedding $E(w)$. We call them three temporary avatars of w, each assigned a different duty.

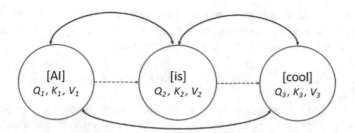

Fig. 10.2 Self-attention between tokens of the sequence "AI is cool." For each token, its query, key, and value are computed using learned functions. Pair-wise dot products are taken between queries and keys. The softmaxed weights are used to take weighted combinations of the values to produce higher-level embeddings of the tokens. This is repeated by several self-attention layers in a transformer. The final output can be post-processed by a task-specific decoder. For example, the decoder might be a classifier which classifies the sentence as having positive sentiment. See Sect. 4.7 to develop intuition about how self-attention performs carving of the input space

They are called the query, value, and key, and they are used to create a new higher-level embedding $E'(w)$ of the word based on self-attention in the context of other words in the input sequence,

$$(E'(w_1), E'(w_2), \ldots, E'(w_N)) = T_1(E(w_1), E(w_2), \ldots, E(w_N)).$$

The positions of the words are also encoded in the input representations, which makes embeddings order-dependent. Positional embeddings can be pre-defined, for example, based on sinusoidal patterns, or they can be learned.

The conventional attention of a target vector h to a source vector v is given by a learned function $\alpha_v = f(h, v)$ or by using the dot product $\alpha_v = h.v$. Then, we scale v by softmaxed α_v to compute the context vector as a weighted combination of different source vectors. In self-attention, it is given by the dot product

$$Q(h)K(v),$$

where Q and K are learned query and key functions, respectively. We scale $V(v)$ by the softmaxed self-attention weight where V is a learned value function. For exact details, see Exercise 26 and its solution [218]. In a sentence,

The jaguar walked on the bank of the Amazon,

self-attention will allow to build representations of the words Jaguar, bank, and Amazon which will encode the fact that they represent an animal beside the Amazon river, rather than a car at a financial bank belonging to a company.

All pairwise bidirectional attentions are computed, and the whole layer T_1 can be viewed as an order-aware generalization of a conventional fully connected layer with layers of complex operations working on vectors produced by the capsules of neurons. There is inherent parallelism which can be used to speed up the computation. The new embedding vectors $E'(w)$ become inputs to the next layer T_2 of the neural network, and the whole process is repeated,

$$(E''(w_1), E''(w_2), \ldots, E''(w_N)) = T_2(E'(w_1), E'(w_2), \ldots, E'(w_N)).$$

Several cascaded self-attention layers yield hierarchical distributed representations of words in a context. A couple of other details are worth mentioning. Skip connections between self-attention layers are used to improve information flow between different levels. There are multiple heads of attention, the results of which are fused back in order to control the computation time. Also, there is a feed-forward layer after every self-attention layer. Due to the pair-wise computation, the transformers have quadratic complexity in the number of tokens, and therefore effort is underway to bring down the time complexity. See [109] in which the quadratic complexity $O(N^2)$ is brought down to $O(N)$, where N is the sequence length, by formulating the self-attention as a linear dot product of kernel feature maps and making use of the associativity property of matrix products. The dot product operation is reformulated as

$$(QK^T)V = \text{similarity}(Q, K)V = \phi(Q)(\phi(K)^T V),$$

where Q, K, and V are, respectively, the query, key, and value representations of the tokens and ϕ is a suitably chosen kernel function.

These representations built by the encoder are input to the decoder just like in any encoder-decoder architecture. Besides this encoder-encoder self-attention, there is encoder-decoder attention and decoder-decoder self-attention. Note that decoder-decoder self-attention has to be unidirectional and causal. See Fig. 10.3.

In a Bidirectional Encoder Representation from Transformer (BERT), a transformer-based architecture is pre-trained on a large text corpus for a self-supervised language modeling task, which entails prediction of masked words in sentences. An example of a masked sentence is

A Bulbul [MASK] is sitting on a [MASK] of the tree,

where the words "bird" and "branch" have been masked out. A classifier is attached to the top of a transformer to predict the masked words. Another task which is used to pre-train is next-sentence prediction. A pair of sentences is given,

[CLS] A Bulbul came to our patio [SEP] Then it flew away,

with a classification token [CLS] and a sentence separator token [SEP]. It is formulated as a binary classification task, with the final highest-level embedding of the [CLS] token as the input to a classifier which outputs whether the second sentence is a valid next sentence or not.

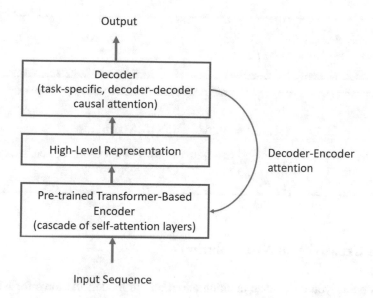

Fig. 10.3 A general framework to solve NLU tasks with an encoder-decoder architecture where the encoder is implemented by a transformer

The training is bidirectional, and the result is such that one gets pre-trained contextual embeddings of words at different layers of the AI model. The embedding of the word "bank" is different when used in the phrase "river bank" than when used in "financial bank." Pre-training facilitates building of customized NLU solutions for different NLU tasks by reusing the same pre-trained encoder and applying transfer learning (fine-tuning). For a new task, a customized decoder is added, and the encoder-decoder architecture is fine-tuned for the given task on domain-specific data. Therefore this is similar to the standard technique of solving computer vision tasks by taking a pre-trained model and fine-tuning it for transfer learning. See [48, 218] for more reading on the subject.

To understand the geometry of BERT, see [177], where it is shown that linguistic features seem to be represented in separate semantic and syntactic subspaces. Though BERT has been trained on tasks which don't involve syntax, it is interesting to note that it learns certain syntactic linguistic features, such as distance between words in a parse tree. A meaning of a word is called word sense. For semantic features, BERT can disambiguate word senses by placing them in separate clusters in the embedding space.

An important application of NLU is in medicine and life sciences research. A large number of biomedical articles are published, which have to be mined for tasks such as biomedical named-entity recognition, biomedical relation extraction, and biomedical question answering. For customization of BERT for biomedical text mining, see [125].

> An encoder-decoder architecture coupled with attention or self-attention provides the basis for AI-based NLU.

> ### ** Exercise 26   Self-Attention
>
> Mathematically formulate how self-attention works in terms of queries, keys, and values; see [218].

10.6   Creativity in NLU Solutions

At this point, you have some intuition about all the modern AI tools for NLU. To solve your problem, you have to creatively come up with a relevant architecture. Here are some examples which will guide you in coming up with your own architecture.

Suppose you want to perform *sentiment classification*, where the sentiment is either positive or negative. Use an encoder-decoder architecture in which the encoder is a bidirectional RNN with LSTM cells. The decoder is very simple, and it can be a couple of fully connected layers with a single classification output. Therefore the architecture is

$$R = E(s),$$
$$y = D(R),$$

where s is the input text sequence, y is the positive sentiment probability, E is a bidirectional RNN, R is the set of hidden states of the encoder, and D is a feed-forward network. Note that R can also be taken as the last hidden state if E is a unidirectional RNN. The loss function is the cross-entropy function.

For the *named-entity recognition* problem, each word is classified as a type of entity. The sentiment classification network above is easily generalized to a word-level multi-class architecture with softmax classification layers.

Consider the problem of generation of text based on *language modeling* starting with an initial token using an RNN which serves as the decoder. There is no need for an encoder. With the initial token as the input, the decoder outputs the probabilities for the words at each stage, from which a word is stochastically chosen (generated). The chosen word becomes the input to the next decoding stage. Therefore, starting with the word "cat," one can generate "is" as the next word, followed by "chasing," "a," and "mouse." Interestingly this modeling can also be done when the tokens are individual characters. Therefore, if w_0 is the initial token, the architecture is

$$y_{t+1} = D(y_t), \, y_0 = w_0,$$

where the recurrent dependence on the hidden states is not explicitly shown. For image captioning, the initial token is the output of an encoder CNN E, that is, $w_0 = E(I)$, where I is the input image.

In some NLU tasks, multiple sequences of tokens have to be processed. For example, in *sentence similarity*, the input is a sentence pair. In the previous chapter on computer vision, we learned that the Siamese CNN architecture provides a solution to the image similarity problem. Therefore, one can consider the Siamese RNN architecture to measure the distance between two sentences,

$$R_1 = E(s_1),$$

$$R_2 = E(s_2),$$

$$d = \text{dist}(R_1, R_2),$$

where s_1 and s_2 are the sentences, E is the RNN encoder, and R_1 and R_2 are the representations of the sentences built by the RNN. If it is the problem of matching a query sentence against a database of sentences, then instead of having the same RNN, one can use two RNNs E_1 and E_2 for s_1 and s_2, respectively. See [146].

At the same time, there is no limit to creativity in NLU. For example, consider the following approach for sentence similarity. Suppose two sentences A and B are "This is a good book on AI" and "This book explains AI well," respectively. The pair is encoded as

[CLS] This is a good book on AI [SEP] This book explains AI well

where [CLS] is the classification token and [SEP] is the separator token. The input sequence is encoded by adding the token embeddings, the position embeddings, and the segment (A or B) embeddings. Therefore, the embedding of the word "good" at the fourth position in the sentence A is

$$E(\text{good}) = E_T(\text{good}) + E_P(4) + E_S(A),$$

where E_T, E_P, and E_S are the learned embeddings for the tokens, positions, and segments, respectively. Recall that a learned embedding is just the first learnable layer which maps a one-hot vector representation of categorical variables to dense representations. The encoded input sequence is input to the first self-attention layer of a transformer. The final output of the transformer is a rich, order-aware, contextual representation E^* of the tokens. For sentence similarity, feed the final representation of the [CLS] token to a feed-forward neural network,

$$y = F(E^*(CLS)),$$

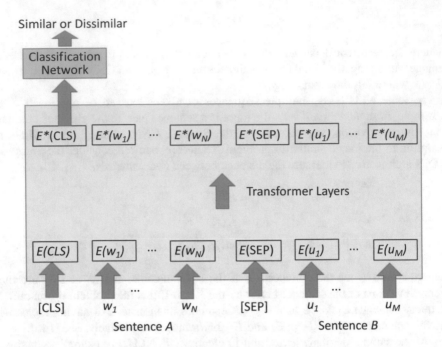

Similar or Dissimilar

Fig. 10.4 A transformer-based architecture for the sentence similarity task. The inputs are sentences A and B along with special tokens [CLS] and [SEP]. The learned embeddings of tokens are the input to the first transformer layer, whose self-attention layers build a rich, order-aware, contextual, higher-level embeddings (representations) of the tokens. The highest-level representation E^*(CLS) of the classification token is input to a classification network. An alternative architecture for sentence similarity is based on the Siamese RNN

where F is the classification network; see Fig. 10.4. The transformer layers can be pre-trained on a denoising language modeling task such as the prediction of masked tokens on a large text corpus. Then, the classification network F is attached, and the whole network is fine-tuned using transfer learning. For details on the above, see [48].

A special case of the above is *single-sentence classification* such as sentiment classification, in which the second sentence B is empty. For word-level classification, such as named-entity recognition or part-of-speech tagging, the classification network F is applied to each output token rather than just on [CLS].

For a *question-answer* system, the sentence A is the question, the sentence B is the paragraph,

[CLS] question tokens [SEP] paragraph tokens

and the output is a sequence of tokens and two special tokens T_S and T_E which encode the start and the end, respectively, of the answer in the output sequence. The starting word of the answer is that output token w for which the similarity, for example, measured with dot product, with T_S is maximized., similarly for the end of the answer. To build a question-answer system against a big corpus, first a search engine has to be developed which will retrieve all candidate passages, which then have to be input to the AI model. Whether to input the retrieved passages individually or all together as one big paragraph will be a system design choice.

By endowing a question-answer system with a conversational memory, one can take first steps toward building a conversational system known as a *chatbot*, in which the goal is to retrieve answers. Building of a chatbot which can hold long, human-like, meaningful conversations is an open problem in NLU. By giving a human-like persona to the chatbot which is a big biographical document of its background, history, identity, and personality, and combining it with a general question-and-answer system against a corpus of commonly known facts and knowledge about the world, the first step toward such a chatbot can be taken, where the goal is just to chat.

For abstractive *text summarization*, the problem is posed as a text-to-text problem [170]. Therefore, one can adopt the same solution as for language translation, in which the decoder produces a summary using the encoder-decoder attention. The architecture is trained under classification loss. For the encoder, the transformer architecture is used. For extreme summarization, in which the goal is to produce one-sentence summary of scientific papers using transformers, see [23], which performs both extractive and abstractive summarization. In today's "too long; didn't read" world, where papers are being written at an ever-increasing pace, this is a useful application of AI.

One implementation issue in NLU systems is that of the variable-sized input. The input is often made of a fixed size by padding the sequence with a [PAD] token. Of course, this won't be needed in AI frameworks such as PyTorch where the computation graph is made dynamically. Even in PyTorch, for minibatches, the inputs have to be padded to the same size. The padding can be done in a pre-padding or a post-padding manner depending on whether it is at the beginning or at the end of the sequence. Pre-padding is preferred when the last hidden state of the encoder RNN is used as the representation of the sequence.

In computer vision, a wide variety of pre-trained CNN models are available which can be fine-tuned for specific tasks. In recent years, a number of NLU models have become available which have been pre-trained on a large corpus. Therefore, for your particular NLU task, the first step should be to create a baseline by using a pre-trained model and fine-tuning it for your own task.

Is the step of fine-tuning absolutely necessary? Consider the following large-scale self-supervised learning strategy.

1. A transformer-based model is pre-trained on a very large corpus with hundreds of billions of tokens on self-supervised language modeling tasks such as next-word generation, masked-word prediction, or next-sentence prediction.
2. The model has hundreds of billions of parameters. This is needed to capture patterns for multiple tasks such as translation, question answering, and summarization, which are naturally occurring parts of a large text corpus.

In the future, extend the sizes of the model and the dataset to trillions. Would the flexibility of language allow the task description to be subsumed by the language and therefore folded into the input, thereby alleviating the need for fine-tuning? That is, suppose you want to translate from English to French. Then your input sequence is

$$\text{Translate English to French: Cheese} \implies \text{[NEXT WORD]},$$

where the assumption is that the AI has seen similar examples for this task in the training set. The above is zero-shot learning. One can supply K examples as part of the input for few-shot learning. Work in [20] known as GPT-3 (generative pre-trained transformer 3) using pre-training for the next-word prediction task seems to suggest that this is a viable approach. What is being claimed is that language modeling leads to a *universal, multi-task, few-shot* learner. In this book, we emphasized that AI builds a representation of the data. In-context conditioning to a particular sub-task builds a representation of the task and a few examples (demonstrations) as a part of the input sequence itself,

$$R = E(t, d, x),$$

$$y = D(R),$$

where E is an encoder which encodes an in-context representation R of a task t, a few demonstrations d, and the data x, which is used by the decoder D to predict the output y. The task description in the above example is "Translate English to French." A demonstration could have been "sea otter => loutre de mer." These are parts of the input itself and enable the model to perform the task through this context without any fine-tuning. Note that d is empty for zero-shot learning.

> *NLU offers opportunities for a lot of creativity in the design of AI solutions; this includes design choices in pre-training, input embeddings, output structure, and architecture. It is a fast-moving application area of AI.*

10.7   AI and Human Culture

NLU works with the latent semantic space of thoughts and concepts, and therefore this space is a reflection of how the human mind thinks and the present state of human societies.

One of the reasons why the human species has been able to progress is that the human mind is able to create *imagined realities* of things and concepts which don't exist anywhere except in the minds of people who agree on the existence of these imaginary things. Concepts such as the Microsoft corporation, British democracy, or the American Dream are imagined realities. They exist in a social and cultural context, because a group of people agree on their meanings and start acting as if these things are real; see [75]. They become so real that battles are fought over them. Over time, new imagined realities get created on top of the existing ones, and the abstraction of our ideas increases. This is analogous to the way a multi-layer transformer builds higher-level concepts based on the underlying concepts built by the preceding layers.

Once these imagined realities get created in the minds of people and they make it into the text documents, then AI models such as transformers will build distributed representations of the new concepts. For the connection between AI and distributed representations, see [87], and refer to Sect. 4.10. The latent semantic space dynamically changes and evolves over time.

What about an arbitrary point in today's semantic space? What does it represent? A point in the representation space may not exist today in the human mind, but maybe tomorrow it will become an imagined reality. A point in the semantic space is a firing pattern of neurons (concepts). Therefore, AI is creating these imagined realities in the activation patterns of neurons, which may or may not materialize in reality. This generation of new concepts is very much like the creation of new images by the AI. This is analogous to how the human mind is able to bring concepts together in a novel way to believe in a new imagined reality, for example, the concept of "ethical AI" becomes as real as any other thing, and ushers in a new era in the way we research, build, and use AI. At the same time, ethical AI of the early twenty-first century is going to be different from what will be meant by the term after 1000 years.

Some newly created ideas facilitate significant cultural evolution because societies design objective functions to be optimized in the pursuit of these new goals. For example, American democracy didn't take root on the very first day. It followed an iterative algorithm in maximizing democratic principles by granting voting rights to everyone over time. There are many historical versions of American democracy. The evolution happens because language allows us to create judgements and opinions, which are really labels or annotations and which guide this iterative progress. Language serves as an annotation tool widely used by the human mind, and debates between people are often about these labels. Language creates imagined realities, annotates them, and drives the cultural evolution. Even the entire social system can be optimized in a space of all possible systems.

Therefore, there is a direct link between the way AI works and the way human culture works. Both work in the abstract *thought space*. The thought space built by AI where NLU entities live will evolve over time, in lockstep with the evolution of human culture and language. It will become closer to the way we organize our abstractions. In the future progress of AI, we should witness new ways of building the thought space.

Thoughts are more complex than D-dimensional vectors. Their actual semantics has to be inferred not just through co-occurrences but with all the underlying semantic structures within an integrated world model. Furthermore, the hypothesis of linguistic relativity, also known as *Sapir–Whorf hypothesis*, states that the way we conceptualize and understand something is related to the way our language expresses it; therefore, semantics gets influenced by the structure of our language. Do words or thoughts come first? Making AI preform NLU is going to help us find the answers to all these challenging questions.

10.8   Recommender Systems

The idea of semantic embedding and distributed representation can be applied to other domains such as recommender systems. Users and products can be represented in a joint space which shows affinity between them. A user who likes a particular product gets placed close to that product.

See Fig. 10.5 for an example of a movie recommendation system. Suppose there are N users and M movies which are embedded in a D-dimensional space where D can be in the hundreds. In word embedding, co-occurrence of words in a text corpus is used. In a recommender system, the co-occurrence gets replaced by the rating value with which the user evaluated the movie.

In the joint space, movie-user interactions are modeled as the inner product. Matrix multiplication computes the inner products of rows of one matrix with columns of another. Therefore, the joint embedding of users and movies is formulated as a matrix factorization technique,

$$R = UV^T,$$

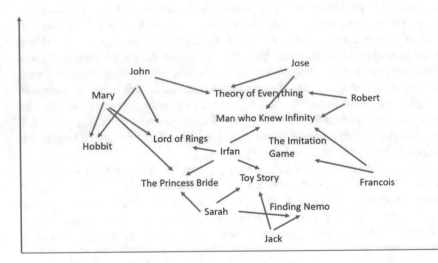

Fig. 10.5 Co-embedding of movies and users in a latent semantic space. The arrows indicate whether a user liked a movie. In collaborative filtering, to make a recommendation of a movie to a user, look at what else the users who like similar movies as the given user have liked. The embeddings and the recommendations can be learned by an AI model

where R is an $N \times M$ rating matrix, U is an $N \times K$ embedding matrix of the users, and V is an $M \times K$ embedding matrix of the movies. Note that R will have missing values.

Singular value decomposition (SVD) provides a matrix factorization of R. A low-dimensional subspace spanned by the K largest singular vectors is retained, and other singular values are discarded. SVD is constrained because it restricts the matrices to be orthogonal, that is, the transformation of the data space is a rotation with scaling. Therefore, users are points in an orthogonal space spanned by movies and vice versa. The SVD-based approach doesn't work well in practice because the rating matrix can be sparse with missing values, and imputation of missing values is expensive and not reliable.

An effective matrix factorization method is that of alternating least squares (ALS), which optimizes a loss function for approximating R by UV^T in a K-dimensional space. For a user-movie interaction, the loss is

$$(r_{u,m} - p_u q_m)^2,$$

where $r_{u,m}$ is a user-movie rating, p_u is the embedding of a user u, and q_m is the embedding of a movie m. Furthermore, the loss function is regularized on the L^2 norm of the embedding vectors. The optimization problem is solved by the gradient descent method, and the ALS method approximates it by using alternating convex optimization steps; see [115].

Once the matrix has been factorized by ALS, an AI model is trained to predict the recommendation score of a user u and a movie m,

$$r_{u,m} = F(p_u, q_m)$$

where F is implemented by a DNN. This generalizes the inner-product-based rating score to a learned function. The embedding layers are initialized to the pre-trained embeddings obtained from ALS. During training, the AI model will fine-tune the ALS embeddings further. At the inference, it predicts the ratings by using the learned function. This approach is called Neural Collaborative Filtering (NCF). See [80] for work in using AI for collaborative filtering.

> *The idea of embedding words in a semantic space can be generalized to recommender systems, in which users and products are represented in a joint space.*

10.9   Reward-Based Formulations

For NLU as well as recommender systems, there are problems in which there is sequential decision-making possibly followed by a reward. This could be a chatbot trying to sell a product or address a technical support problem. The sequence of actions could be a series of clicks and steps taken by a user which will determine the final recommendation for a movie. For each step, we want to predict what the user is likely to do next and what should be the action taken by the interacting AI model. It is natural to formulate such problems in terms of reinforcement learning. This is one of the next frontiers in NLU, and we should expect to see progress in combining NLU with RL.

10.10   Chapter Summary

AI for NLU starts with distributed representations or embedding of words. Glove and word2vec are commonly used word embeddings. They capture semantic similarity of words in a high-dimensional space of latent concepts. The first layer of an AI model for NLU is an embedding layer which converts one-hot representation of words into embeddings and which can be initialized by pre-trained embeddings. For sequence-to-sequence tasks such as machine translation, the encoder-decoder architecture with the attention mechanism has proved to be effective. The encoder encodes the input into a distributed representation which is used by the decoder to produce a sequential output. The attention mechanism allows the decoder to refer back to the states of the encoder. The attention mechanism is generalized to self-attention in which the encoder builds a contextual representation of the input by computing pairwise attention scores of words within a sentence. A transformer uses multiple self-attention layers to build a hierarchical representation of a token sequence, which resembles the way the human mind builds more abstract concepts. Recommender systems derive inspiration from the idea of semantic embedding. Therefore, users and products can be embedded in a joint space, and an AI model for a recommender system uses the joint embeddings for prediction tasks, as in NLU. Alternating least squares can be used to pre-train these user-product embeddings.

Chapter 11
Lend Me Your Ear

Before I lost my voice, it was slurred, so only those close to me could understand, but with the computer voice, I found I could give popular lectures. I enjoy communicating science. It is important that the public understands basic science, if they are not to leave vital decisions to others.

Stephen Hawking

I cannot speak well enough to be unintelligible.

Jane Austen

Communication is key to human experience. Evolution enabled living things to pick up vibrations in air to make sense of what is happening around them and to proactively produce such vibrations in order to interact with others. Humans have taken this gift of sound recognition and synthesis to the next level by combining it with language and music, resulting in conversations, songs, symphonies, lectures, debates, and movie dialogues. It would be fantastic if AI were to receive this gift from us. This chapter describes how you can create this ability in software. In fact, ASR was the first major industrial application of AI; see [91] for phonetic classification using deep learning.

S. Dube, *An Intuitive Exploration of Artificial Intelligence*,
https://doi.org/10.1007/978-3-030-68624-6_11

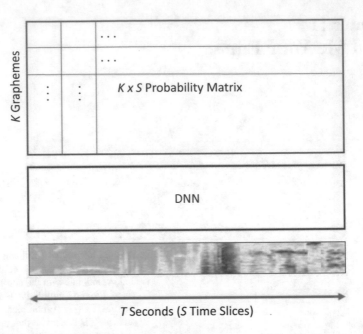

Fig. 11.1 Connectionist temporal classification for speech recognition. (The shown spectrogram image is from the sensors of NASA's Cassini spacecraft, as it crossed through Saturn's D ring on May 28, 2017. Courtesy NASA/JPL-Caltech/University of Iowa)

ASR can be broken into two stages: (1) acoustic modeling and (2) language modeling. Acoustic modeling takes a speech waveform as the input and breaks it into phonemes, the fundamental units of spoken sound. Language modeling takes phonemes as input and produces words according to the language constraints and vocabulary.

The starting point for both classical ASR and AI-based ASR is the spectrogram. A spectrogram shows the variation of a power spectrum with time divided into small units (say, 10 milliseconds) called frames. Frames are on the x-axis and the frequency on the y-axis. The power is visually shown by greyness. See Fig. 11.1, in which a spectrogram is shown with a heat map. In Sect. 9.7 we assisted AI for video analysis by precomputing the optical flow, and in the same way, we assist AI in ASR by precomputing the spectrogram.

In speech synthesis, we go in the other direction. The input is a piece of text, and the output is a spectrogram and therefore a synthesized voice. In this chapter, we will build conceptual understanding of speech recognition as well as speech synthesis.

11.1 Classical Speech Recognition

Classical ASR systems perform mel scale filter bank analysis and derive cepstral coefficients called Mel Frequency Cepstral Coefficients (MFCCs). The mel scale is a non-linear spectral scale reflecting how the human ear works. You can distinguish 400 Hz from 450 Hz better than 4000 Hz from 4050 Hz. It is a log scale. The cepstrum, an anagram of spectrum, is the Fourier transform of the log power spectrum and therefore referred to as the spectrum of a spectrum. Consider MFCCs as acoustic features hand-engineered from the spectrogram. A phoneme can be visually traced as a temporal trajectory of broad strong peaks, known as formants, in the spectrogram. The formant trajectory is divided into segments. To recognize these trajectories, Mel Frequency Cepstral Coefficients (MFCC)s are fed into an HMM whose states correspond to formant segments. The HMM outputs the most likely sequences of phonemes based on the observations of MFCC features.

Language models combine phonemes into words, which are then combined to form sentences. Therefore one can state with certainty that a hierarchy of HMMs must have been tried in the past; see [52]. Sentence-level HMM, word-level HMM, and phoneme-level HMM are natural ideas in classical ASR. Language modeling brings in several linguistic tools and techniques, such as vocabulary, lexical analysis, N-gram models, statistical models, and word transition graphs.

11.2 Spectrogram to Transcription

Let us now move to the exciting topic of AI-based ASR. For acoustic modeling, the idea is to input a spectrogram directly into a DNN and get the transcription, which is a sequence of phonemes or graphemes (characters). No explicit computation of features is performed, and it is an example of automated feature engineering by AI.

11.2.1 Alignment-Free Temporal Connections

Let us say you utter the word "Awesome" in multiple ways varying your tone, speed, and pronunciation. Each utterance yields a spectrogram. The goal is to produce "Awesome" given a spectrogram, that is, we want to output the symbol sequence,

$$['A', 'W', 'E', 'S', 'O', 'M', 'E'] = F(S),$$

where S is the spectrogram of "Awesome." The algorithm has to be robust to different ways the same word can be spoken. One can speak slowly or fast. There is a stretching or shrinking of the spectrogram. Therefore, instead of building one function F which maps the spectrogram to the word w,

$$w = F(S),$$

we will first build a stretched or contracted intermediate representation M from which the word will be produced,

$$M = F_1(S),$$
$$w = F_2(M).$$

This is needed because a 3-second-long utterance will have three times as many frames as 1-second-long utterance, but the output transcription will be of the same length, that is, seven characters. The output symbols and input segments have variable alignment. In order to solve the alignment problem, a K-dimensional probability vector is produced for K phonemes (or graphemes) for each input frame using a DNN. The DNN may optionally consist of a few initial convolutional layers followed by RNN layers which are bidirectional since we are given full temporal context. For any i-th segment of the spectrogram S, the output of the model is a probability vector for the segment

$$[p_A, p_B, p_C, \ldots, p_Y, p_Z, p_{\text{Blank}}, p_{\text{Space}}] = F(S, i),$$

where "Blank" (for no label) is a special symbol which can be squeezed out of the transcription and "Space" is a pause in time. If a frame is 20 milliseconds long, then for a 3-second utterance, we have a $K \times 150$ output probability matrix, and for a 1-second utterance, we get $K \times 50$. We call this the probability or score matrix M.

In order to get "Awesome" from M, the first thing an AI engineer can do is to visualize M and see whether "Awesome" can be read off M when traversing from left to right, squeezing out the blanks. To do it automatically, a non-trivial search problem is solved because M leads to several competing transcriptions besides "Awesome" which have to be eliminated. A candidate transcription is a path going from left to right, in which blanks can be deleted. By bringing in tools such as dynamic programming, HMM, and N-gram models from language modeling and beam search, one can infer the correct transcription. The exposition in this section is inspired by Connectionist Temporal Classification (CTC); see [65, 68]. CTC can be viewed as an output layer for a variable-length input which outputs the probabilities of different labels including a special symbol of blank, which can just be deleted. See Fig. 11.1. See the solution of the following exercise to learn how the best transcription is selected from the candidate transcriptions.

** Exercise 27   CTC

Mathematically formulate how transcriptions can be inferred from the probability matrix M.

CTC is an output layer which produces a probability matrix as the output of an AI model which takes a speech spectrogram as the input. This probability matrix is further processed to produce the transcription. A special symbol of blank (no label) is used to deal with the variable-length input.

11.2.2 End-to-End Solution

In this section, we develop intuition behind an end-to-end solution for acoustic modeling based on Fig. 3.5. The model is based on the encoder-decoder architecture and the attention mechanism; see Sect. 3.2.4.

The first step is to create an embedding of the input spectrogram with the encoder. The encoder is a CNN or a bidirectional RNN or a combination of the two, which takes the spectrogram frames as the input and creates an internal vector representation R. The second step is to employ a Seq2Seq model, for example, a multi-layer LSTM with the attention mechanism as the decoder. Recall that the attention memory implements memory of hidden states of the encoder to be selectively accessed or attended to at every decoding time step.

The symbol "Begin" and R are fed to the decoder F to start the decoding process. The output is a K-dimensional probability vector which hopefully has maximum (or near maximum) value for grapheme "A," that is,

$$`A' = F(`Begin', R, 0),$$

where 0 means that attention is not being employed in the first time step. If "A" does not have maximum probability, then the hope is that in the beam search "Awesome" will still have maximum aggregate probability. At the next step, an attention-weighting vector is produced which shows the similarity of the current hidden state h of the decoder with the contents of the attention memory. A weighted linear combination of the memory contents yields the context vector c. The decoder concatenates "A," R, and c as the input and produces the next output, hopefully with a maximum (or near maximum) occurring at "W," that is,

$$`W' = F(`A', R, c),$$

where dependence on the previous hidden state is not shown for simplicity. Continuing further, the rest of the characters "E," "S, " "O," "M," and "E" are output followed by a special symbol "End" to terminate the decoding.

The whole system is trained end to end. Note that we may have to apply beam search on the output to select the best overall transcription. This section is inspired by a large amount of recent work in ASR using Seq2Seq models. See [27, 33, 65] as relevant references.

It is natural to ask two questions here. Can the raw waveform rather than the spectrogram be used as an input? Can one use a fully convolutional approach? The answer is affirmative. Consider the following end-to-end approach.

1. Use a front-end CNN to convert the raw waveform into an internal spectrogram-like representation.
2. Use a CNN for acoustic modeling as in CTC.
3. Use a CNN for language modeling which scores candidate transcriptions. In addition, perform beam search on the output of the acoustic modeler.

This end-to-end solution works well and can achieve competitive results; see [250] for further reading.

11.2.3   Don't Listen to Others

At the present time, voice assistant systems have to solve the *cocktail party problem*. In the future, a robot waiter will take your orders in a noisy loud restaurant. The ability to listen to exactly one particular speaker while ignoring other sounds and voices is critical for deployment of ASR in the real world. This is called anchored speech recognition; see [138, 226, 255]. The person who the ASR system needs to listen to is called the anchor. He/she is the one who has either initiated the conversation in a restaurant or spoken the wake word for a personal assistant device. All other speakers should be ignored.

Anchored speech recognition can be solved by AI in the encoder-decoder architecture form as follows. Characteristics of the anchor's voice in the form of an internal representation R are computed,

$$R = E(\text{Anchor's Voice}),$$

by an AI model E, which is the encoder. The representation R is used for detection and rejection of other voices by a second DNN D,

$$y = D(x, R),$$

where x is a speech segment to be accepted or rejected. The decoder model D is trained to classify speech segments into two classes: (1) if a speech segment belongs to the anchor or (2) if the speaker is someone else or noise. The decoder takes R as an additional input at every time step along with a speech segment. The encoder and the decoder are trained together to form an E2E solution.

11.3 Speech Synthesis

Two-way communication is preferable to the one-way case, when just one person talks and the other listens. To enable the AI to respond to us by using speech, the AI has to be given a voice. Given a sentence, we need to produce an audio waveform. In this section, we build insight into speech synthesis (text-to-speech conversion) techniques.

In the *concatenative* approach, a database of recorded speeches is employed,

$$v = \text{Database}(w),$$

where w is the target word and v is its waveform. It is called the concatenative approach because waveforms of short speech segments by the same speaker are combined for utterances of longer sentences.

This method works fine, but the output will lack the naturalness of normal speech because the full context around w is lacking. The way we say words depends greatly on the context. "Awesome" is spoken differently in "You are so unbelievably awesome, my dearest!" and in "We were so disappointed in the product despite a few awesome customer reviews." Acoustic parameters such as pitch, duration, spectrum, and intensity will determine the style of speaking. Phonetic, linguistic, and semantic features of the words in a sentence are computed based on text analysis, and then they are input into a DNN F which outputs the desired acoustic parameters,

$$z = F(g(S), w),$$

where z is the vector of acoustic parameters, w is the target word in the context of the input sentence S, and g computes the text-analysis-based features of the words in S. The DNN is a bidirectional RNN and is trained on a large speech database which has speech segments spoken in different contexts by a person whose voice will become that of the robot or the AI model. Waveforms in the database for w matching the desired parameters are found and used in the robot's voice, that is,

$$v = \text{Database}(w, z),$$

where v is the retrieved waveform.

If one is unable to find a v which is close enough to the desired acoustic parameters, then the method can be complemented by a vocal model G which modifies a recorded voice according to the desired acoustic parameters,

$$v' = \text{Database}(w, z),$$
$$v = G(v', z),$$

where v' is the best recorded voice for w in the database. Therefore the above approach is a concatenative approach guided by a *parametric* method with the fallback option of using a vocal model to make the output sound more natural. The vocal model G is based on signal processing.

Therefore, in general, there are three stages for speech synthesis: (1) text analysis, (2) generation of acoustic or speech parameters, and (3) speech synthesis. Stage 2 is implemented by an AI model. Speech synthesis is aided by having a database of recorded speech segments as in the above approach, and the question is whether one can eliminate the database completely. This is indeed possible, and the speech can be completely synthesized by a vocoder (speech synthesizer) based on the acoustic parameters, making the method completely parametric. The next question is whether one can eliminate the feature-based speech synthesizer and replace it with an AI model. And finally, the most ambitious goal is to build an end-to-end solution which optimizes all the components together, which includes the front-end text analyzer.

AI-based E2E models produce the output waveform or spectrogram directly from the features of the words, $v = F(g(S))$. With continuing progress in speech synthesis, parametric methods and end-to-end solutions are gaining ground over concatenative methods. For advanced methods, Seq2Seq models with the attention mechanism are used which take a sequence of graphemes as the input, build an internal representation R with an encoder, and make a recurrent decoder consume R to output the desired spectrogram; see Sect. 3.2.4. That is, we have the E2E architecture,

$$R = E(S),$$
$$S' = D(R),$$

where S is the input sentence of graphemes (characters), S' is the output spectrogram, E is the encoder, D is the decoder, and R is a representation built by the encoder. See Fig. 11.2. An alternative method creatively uses CNN to implement text-to-speech systems in which causal convolutions, which can be dilated to increase the receptive field of the neurons, are used to map the text features to the raw waveform. These approaches lead to large AI models, and therefore research is underway to reduce the computation time, for both training and inference.

See [24, 156, 165, 195, 225, 252], which have inspired this section, to learn more about the exciting application of AI to speech synthesis. The state of the art is still not at a point which could be called perfect, natural pronunciation, and therefore there is an opportunity to experiment with novel approaches.

Speech recognition and speech synthesis are two exciting applications of AI. Encoder-decoder architectures for sequence-to-sequence modeling coupled with the attention mechanism provide end-to-end solutions for translating a spectrogram into a sequence of graphemes and vice versa. There is an opportunity here to be creative with the architectures with a number of variations which are possible in order to advance the state of the art.

Fig. 11.2 An end-to-end solution for speech synthesis. The input is the sequence of graphemes "AI is cool," and the AI model outputs a spectrogram from which the raw waveform can be obtained

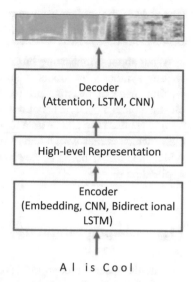

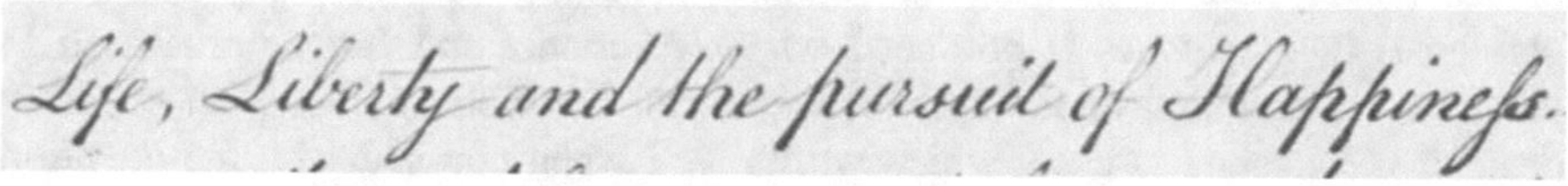

AI is Cool

Fig. 11.3 Cursive handwriting recognition can be performed by AI using approaches similar to those for speech recognition. It is an image-to-grapheme transcription problem. (The image is a part of the United States Declaration of Independence, 4 July 1776)

11.4 Handwriting Recognition

Handwriting recognition fits well with the framework outlined in this chapter. Instead of a spectrogram, a 2D pixel image of a handwritten sentence is input to an encoder-decoder architecture with the attention mechanism, where for the encoder it is natural to use a CNN. That is, we have the architecture

$$R = E(I),$$
$$s = D(R),$$

where I is the input image, s is the output sentence, E is the encoder, D is the decoder, and R is a representation built by the encoder. See Fig. 11.3.

Alternatively, the CTC approach can be used,

$$M = E(I),$$
$$s = F(M),$$

where M is the probability score matrix output by an AI model E with the CTC output layer, which is post-processed to produce the sentence. For further reading, see [67, 166].

What about synthesizing human-like handwriting for a given sentence s? Handwriting synthesis by robots is an oxymoron because machines use perfect handwriting made of machine-printed words on a computer screen. In the future, an old-fashioned, nostalgic robot will be able to personalize its output text with its own unique handwriting and be able to write letters to us. In the meantime, a touch screen tablet which comes with a handwriting pen seems like a great buy. Combined with ASR, such an AI-based device is a very ergonomic way of working and is a fine application of AI.

11.5   Chapter Summary

Automatic speech recognition (ASR) was the first major industrial application of AI. For ASR, CTC is used for acoustic modeling. A probability matrix for phonemes or graphemes is produced by an AI model and further processed to yield the transcription of the speech waveform. For E2E solutions, an encoder-decoder architecture is used, which works well when coupled with the attention mechanism. Anchored speech recognition can be solved by an encoder-decoder architecture; it classifies a speech segment either as belonging to the anchor or not by encoding the anchor's speech in a representation. Speech synthesis or text-to-speech transcription can be solved with the help of a database which is searched with the help of an AI model which determines the desired acoustic parameters for a word in a given context. Another alternative for speech synthesis is an encoder-decoder architecture in which the decoder outputs a spectrogram. Finally, cursive handwriting recognition is an image-to-text transcription problem and is amenable to techniques very similar to the ones used for the spectrogram-to-text transcription problem.

Chapter 12
Create Your Shire and Rivendell

> You don't take a photograph, you make it.
>
> Ansel Adams

> If real is what you can feel, smell, taste and see, then 'real' is simply electrical signals interpreted by your brain.
>
> Morpheus in *Matrix*

One of the joys of Hollywood movies is stepping into worlds, dreamlike and fantastic. The Shire and Rivendell are magical works of art created by the human mind. Paintings, theater, movies, music, stories, poetry, and myths express the human mind's innate desire to express itself. When we look at the cave paintings of Chauvet Cave, we need no further proof of the intelligence of their creators. We explore the topic of creation in this chapter. Can AI help us in our artistic endeavors? Can it create aesthetically appealing worlds? Can it add special effects to images and videos?

In Hollywood, computer-generated imagery (CGI) has become indispensable, and its effective use has helped in creating Oscar-winning masterpieces. In order to make CGI more realistic, real-world input is used in the synthesis. Behind Gollum is Andy Serkis. The computer graphical technique of subsurface scattering realistically models reflection of light on and under Gollum's skin, so that you can even see his veins. When Gollum imitates Andy Serkis's real-world movements and facial expressions, and leads Frodo and Sam to Shelob on the steep cliffs overlooking Mordor, a transfixing movie experience is created.

AI for computer graphics is based on a key observation that AI is trained on real-world images. Somewhere in the neural connections, there is a model which captures nuances of reality with its colors, shapes, shades, and textures. In order to

Fig. 12.1 Synthetic images of the monarch butterfly and the bulbul bird randomly generated by a large-scale GAN known as BigGAN. BigGAN is a state-of-the-art conditional GAN in which the input is augmented by the class label, and it is trained on 1,000 ImageNet classes. The images were created by the author in 2020 using pretrained models, which were created by DeepMind and released to the public domain

create new worlds, the real world hidden inside the black box has to be brought out in new forms. See Fig. 12.1 for an example of the state of the art.

12.1   From Neurons to Art

12.1.1   DeepDream

We conjectured that there is a version of the real world hidden in the parameters of a DNN. Let's make an attempt to bring it out in the open. Our effort is more like probing the mind of someone. The network is not optimized to create new content; it has been trained, and there is no further training to make the output artistic. We just want to know what real world is hidden in it. To see how someone's mind works, a strategy is to provide stimuli to the brain and observe how it responds. The same strategy will work for AI.

Suppose a psychologist shows you a photograph and probes the strength of neural firings in your brain. He/she tweaks the photograph in a way which enhances the firings of these neurons further in order to understand which patterns in the photo the neurons respond to the most. If in the tweaked photo, flowers become more pronounced or start appearing in the sky, then flowers are buried in your subconscious mind. Because of the positively reinforcing feedback loop, soon the photograph will be covered with flowers. The motivating idea is that if there is an activation vector of the neurons, for example,

$$[5, 0, 6, 0, 0, 0, 12, 0, 0, 0],$$

then it should be strengthened further, for example, to

$$[15, 0, 18, 0, 0, 0, 36, 0, 0, 0].$$

Recall that the output of any neuron is a function of the input and learnable parameters:

$$y = F(x, \theta).$$

In the backpropagation algorithm, we have weights as variables to be tweaked and the input image pixels as constants. We can do the opposite. That is, keep the weights frozen and have the pixels as variables. This allows us to see how the network will respond if we change the stimulus.

Here is the proposed algorithm. Choose a layer L of the AI model you want to probe. Look at the activation outputs of the neurons in that layer. Set the sum of these values to be the objective function $O(X)$ and maximize with the gradient *ascent* algorithm, that is,

$$O(X) = \sum_{i \in L} F_i(X, \theta),$$

where X is the input image and F_i is the activation of the i-th neuron in the target layer L. Run backpropagation all the way to the input layer, differentiating $O(X)$ with respect to the input pixels X. Tweak the pixels in the direction of increasing $O(X)$ by taking a gradient ascent step. Keep on climbing up the optimization landscape by taking many such steps.

Dreamlike images will appear. Dogs may appear in the pretrained ImageNet CNNs because dog images are widely available on the Internet and get consumed to train AI for the ImageNet competitions. Is this art, dreaming, or just hallucination? You are the best judge. See Fig. 12.2, in which a few iterations of DeepDream are executed.

12.1.2   Style Transfer

An image can be represented in many ways. Frequency-based representation of images using Fourier transform is well-known. Consider Albert Einstein's picture and that of Marilyn Monroe. By combining low frequencies of the former with high

Fig. 12.2 A DeepDream
image, created by the author
(2020)

frequencies of the latter, one makes a nice demonstration. From a distance, human
vision sees low frequencies, making the picture that of Albert Einstein. When one
starts walking closer, the picture will morph into Marilyn Monroe as the human
visual system starts noticing high frequencies and ignoring low frequencies. The
two pictures have been merged into one.

AI can be employed to merge two images to create a third one. The contents of
one picture can be mixed with the style of another using a pretrained network, which
will remain frozen. Call the two images A and B. The idea is to create a third image
Z such that

$$Z = \text{Content}(A) + \text{Style}(B).$$

The activation values of the neurons of a target layer J of A represent its content.
The feature correlation statistics of B represents its style, where we typically use
all the layers. Starting with any input image X (it can be A), iteratively modify it

by running the backpropagation algorithm all the way to the input layer just like in
DeepDream. The objective function to be minimized is the departure from the target
content and the target style,

$$L(X) = \alpha d_1(C_X, C_A) + \beta d_2(S_X, S_B),$$

where C_X and S_X are the content and the style of X, respectively; C_A is the content
of A, and S_B is the style of B; and d_1 and d_2 are suitably chosen distance metrics,
for example, the Euclidean distance or the cosine distance. The total loss function
$L(X)$ is a weighted combination of the two distances given by the hyperparameters
α and β. The gradient descent steps are taken to minimize the total loss.

The style captures how different feature maps correlate with each other. Consider
a feature blob of dimensions $W \times H \times K$ computed by the CNN. The spatial
dimensions are $W \times H$, and there are K feature maps (channels). Reshape the blob
to a matrix M with K rows and $W \times H$ columns. The style matrix is the $K \times K$
covariance matrix:

$$S = MM^T.$$

Matrix multiplication can be viewed in two equivalent ways: (1) pairwise inner
products of the rows of M with the columns of M^T or (2) pairwise outer products
of the columns of M followed by a sum. Either way you are computing correlations
between K vectorized feature channels. This is done for all layers.

The intuition behind the math above is simple. For content transfer, if C_A is the
activation vector,

$$[5, 0, 6, 0, 0, 0, 12, 0, 0, 0],$$

for the target layer J, then C_X is made the same. For style transfer, consider two
feature maps (channel-wise activation vectors) p_B and q_B in some layer of the style
image B,

$$p_B = [20, 0, 2, 10, 15, 0, 30],$$

$$q_B = [14, 0, 3, 11, 20, 0, 35],$$

which have strong positive correlation. Initially, suppose the same feature maps for
X are the following:

$$p_X = [0, 40, 20, 10, 0, 0, 0],$$

$$q_X = [3, 5, 4, 2, 0, 0, 3],$$

which have weaker positive correlation. The goal is to modify them so that the
correlation becomes stronger. That is,

Fig. 12.3 Transferring the style of a dolphin painting to a photograph of flowers. See Sect. 12.1.2. (Credit for the Dolphin Painting (2017): courtesy Saundarya Nair, Grade 6 student. Image created by the author (2020))

$$\text{Correlation}(p_B, q_B) \approx \text{Correlation}(p_X, q_X).$$

Starting with A as your photograph and B as a painting by Picasso, Vincent van Gogh, or Monet, you will have their style transferred to your photograph. Thanks to AI, we now have the talents of artistic geniuses available to us.

See Fig. 12.3, for transfer of the style of a dolphin painting to a photograph of flowers. The dolphin painting has a style, in which there is a strong correlation between the green and blue color blobs and the oriented black contours and lines. There are more blue blobs than green ones. This style is transferred to the flower picture. The oriented structures in the leaves get enhanced. In the pillars, we see oriented streaks. An explanation could be that the gray color of the pillars includes the blue color and the shadows on the pillars are creating oriented structures.

You can read about Neural Art Style Transfer in [56].

12.2 Image Translation

Both DeepDream and neural style transfer can be viewed as very special cases of image translation, which use frozen pretrained models. The techniques are based on the modification of the input.

For general image translation, supervised approaches which modify the AI models are used. Many useful problems can be formulated in terms of image transformation. From paintings, photographs can be obtained and vice versa. Photographs can be enhanced by increasing focus on the foreground and by blurring the background. Noise and artifacts can be removed from photos. A nighttime photograph can be converted into a daytime photograph and vice versa. In medicine, a CT scan can be converted into an MRI scan. A CBCT scan, which can be obtained using faster and cheaper machines, can be converted into a CT scan with higher contrast. In medical radiotherapy, given a 3D segmentation of the target organ and of organs at risk, an AI model can output the 3D dose distribution. In fact, an ideal end-to-end solution will work on the CT scan as the input with the output being the dose distribution. Semantic segmentation is an image translation task, in which an image is translated to a semantic map. Semantic labels can be translated back to images. From sketches, paintings can be obtained and vice versa. From painted historical portraits, photographs can be obtained. From street maps of cities, their aerial images can be obtained and vice versa. Colors can be added to photos and videos, and missing pixels can be inpainted. The resolution of an image can be increased. Given a segment of video frames, the next frame can be predicted.

The ground truth for the paired image translation task is a set of N image translation pairs $(x_1, y_1), (x_2, y_2), \ldots, (x_N, y_N)$. Image translation can be performed by the standard autoencoder. The loss for a translation pair (x, y) to train a translator autoencoder F is the distance

$$d(y, F(x)),$$

which generalizes the reconstruction loss

$$d(x, F(x)).$$

The distance metric d can be the L^2 distance. It has been observed that the use of the L^1 distance results in less blurred images.

Autoencoders within the GAN framework can be used to solve the image translation problem; see [101] for an example. Such an autoencoder-based GAN is called a translator GAN or an autoencoder GAN. The standard GAN, which is defined as a mapping from the generative parameters (random vector) θ to image x, $G(\theta) = x$, is generalized to an encoder-decoder architecture:

$$y = G(x) = D(E(x)).$$

A translator GAN is a special case of *conditional* GANs. The output image is conditioned on additional input, which is the input image for a translator GAN. A class-conditional GAN is another special case in which the class label is the additional input, where the class label is provided as a one-hot vector and goes through an embedding layer; this provides an additional control over the desired class of images we wish to generate at the inference time. In general, for an additional control over the generation process, the input of a generative AI model can be augmented by any desired attributes about the target image, which needs to be generated.

Since translation can often work in both directions, one can train a reverse translator $x = F'(y)$. Both F and F' can be trained together with a combination of loss functions which are either reconstruction based or cyclic:

$$d(y, F(x)), \quad d(x, F'(y)), \quad d(x, F'(F(x))), \quad d(y, F(F'(y))).$$

See work on the cyclic GAN for cycle consistency ideas [257]. Furthermore, see Exercise 28. The exercise dives deeper into how to construct the loss functions for image translation tasks. We consider both paired and unpaired translation tasks.

** Exercise 28   Image Translation

See [101, 257]. Write the loss functions for the discriminator and the generator for the translator GAN. Furthermore, write the loss functions for the cyclic GAN.

12.3   DeepFake

The media is abuzz about DeepFake, the generation of fake content by AI, and examples of fake speech, fake images, and fake videos can be easily found on the Internet. Generative AI learns generative parameters of objects of interest such as cats:

$$g : \Theta \rightarrow X_{\text{Cats}}.$$

Therefore, it can generate new content, that is, new cats. Instead of a general class of objects such as cats, a class for a specific person B can be made the target class for training,

$$g : \Theta \rightarrow X_B,$$

where X_B is the set of all possible variations of B. The trained AI model can generate infinite variations of B. To generate specific fake content, one can sample desired parameters for the new content,

$$g(\theta_{\text{target}}).$$

In DeepFake videos on the Internet, there are examples in which the speech of one person is changed to somebody else's voice and the face of a person in a video is swapped for another face. By using the generative parameters from an instance of another person A, the facial expressions and the speech of A in that instance can be transferred to an instance of B, provided the parameters have the same semantic interpretation. Therefore, one first generates the desired parameters using a learned function h,

$$\theta_{target} = h(A),$$

which are then used in the function g to generate a fake instance of B.

Let's develop intuition behind this straightforward technique further by using autoencoders. Train an autoencoder on face images, such that we have the same encoder E for all faces but a separate decoder for each face. Having the same encoder ensures that the generative parameters have the *same semantic interpretation*. The encoder E will output a vector of the face generative parameters θ. Take images A and B. Then, to transfer attributes of A to B, apply the following:

$$\theta_{target} = E(A),$$
$$B' = D_B(\theta),$$

where D_B is the decoder for B. Therefore, if A is smiling, then a smile will be added to B. Train another autoencoder on speech segments in a similar way. Have the same encoder for all speakers but a separate decoder for each. Transfer the speech attributes, representing phonemes or graphemes, of A to B. If A is saying "Hello, world," then B will say the same. In this way the attributes of a video of A can be transferred to B, who will say and do what A was saying and doing.

Such a technology does have positive applications. For example, personalized multilingual videos can be generated for educational purposes using DeepFake.

See [1, 152, 213] to learn more about DeepFake.

12.4   Creative Applications

Children have great fun when they paint, sketch, and draw. With generative AI, we can all afford to experience moments of creativity. Some of these can be practically useful and lead to new applications. We give a few examples.

See Fig. 12.1 for synthetic images of monarch butterflies and bulbul birds using BigGAN; see [19]. It is a class-conditional GAN in which the GAN is conditioned on all 1,000 ImageNet categories. The number of parameters is very large, in a few hundreds of millions. Compared with the prior art, it is an impressive generative AI model; though for certain difficult categories involving people, further improvement needs to be made. For further improvement of BigGAN by combining it with a bidirectional GAN, which attaches an encoder to GAN, see [49].

Simulation is a big application area of generative AI. In computer graphical games such as the classic Pac-Man, in which the player munches dots in a maze while avoiding ghosts, generative AI can be applied creatively to render new screen frames given the key pressed by the user; see [111].

In the textile industry, whole-garment knitting machines are revolutionizing mass customization. Programming such a machine is very tedious. Instructions in existing pattern books are geared towards hand knitting. It will be convenient to automatically produce these machine-level knitting instructions, given a desired piece of fabric. In [108], the inverse design problem of translating fabric to machine knitting is solved with the help of generative AI.

With the world moving towards online shopping, it is natural to see whether one can try out items of apparel virtually before purchasing them. Generative AI can be employed to build a solution. One can bring in the tools of semantic segmentation, pose estimation, autoencoders, GANs, and style transfer to build a combinatorial solution. See [148] for an example solution, which combines such components.

> Generative AI, by virtue of being trained on real-world data, is capable of generating realistic-looking content. It can be trained not only to generate new content belonging to a class but as a tool to transform a given content.

12.5   Chapter Summary

Pretrained AI models can be used to tweak an input image to translate it into DeepDream by strengthening the firings of neurons. Neural art style transfer consists of tweaking the input image so that its feature blobs match with those of a content image as well as its statistics match those of a style image. DeepDream and style transfer do not require any training, and everything is done by modifying the input using the backpropagation algorithm. For general image translation, a standard autoencoder or an autoencoder GAN can be trained on paired images or paired image sets. Any generative AI can generate fake content by sampling the latent space of the generative parameters. A specific case of DeepFake which swaps the attributes of a video of a person onto that of another person can be accomplished by obtaining the generative parameters from the first person and decoding them for the second person. There is no limit to creativity in generative AI both for fun or for useful applications.

Chapter 13
Math to Code to Petaflops

I started Keras for my own use, pretty much ... I open-sourced it, and it grew from there ... what made it different at the time was this: it was pretty accessible and easy to use compared to other options, it supported both RNNs and CNNs (that was a first, I believe), and it made you define models via Python code rather than via configuration files (which had been the most popular approach previously).

François Chollet

It's a foregone conclusion that we're going to see some really gigantic AI models because of the creation of Ampere and this GPU generation.

Jensen Huang

AI is a great example of how rapid advances in software tools and hardware have synergized to create a revolution in computing. High-level tools which are built on others allow higher abstraction and quick experimentation, making it easy for AI engineers to forge ahead. The speed of execution of brilliant ideas matters. When we look back at the published literature of the past, one can see that people have been very smart all the time. It was lack of tools and technology, which prevented them from building things in their time.

In this chapter, we develop an insight into how the mathematics behind AI translates into software, which runs at a blazing speed in the silicon.

S. Dube, *An Intuitive Exploration of Artificial Intelligence*, https://doi.org/10.1007/978-3-030-68624-6_13

13.1   Software Frameworks

The science behind AI is described by mathematics, which takes concrete shape in the form of software, and we enter the world of design choices, software tools, data handling, and hardware issues. In this arena, experiments are done.

13.1.1   The Twentieth Century

Ideas in AI have to be meticulously turned into a software program. Imagine writing it all in C/C++ from scratch if we were back in the 1970s and 1980s. In fact, Caffe, one of the first AI open-source tools, was written in C++. Therefore, such work can be done by people with excellent coding skills, and at the same time, ideally one needs higher-level libraries to expedite prototyping.

Realizing the limitations of C/C++ for research work, scientific programming languages were developed in the 1970s and 1980s, and R, Matlab, and IDL gradually became popular due to their high-level programming features and excellent visualization tools. In these frameworks, one could work with matrices and data arrays easily and had access to a large suite of libraries of useful functions. For symbolic mathematics, Mathematica was developed. Then came the 1990s and the first decade of the twenty-first century, and in industry, companies adopted Microsoft tools with the rapid rise of Microsoft Visual Studio. For the front-end work, Java took over the world.

13.1.2   The Twenty-First Century

This was to change soon. Unix made a comeback in new avatars of Linux, Debian, and Ubuntu, and Mac OS became Linux based. Linux started conquering industry as developers adopted it, and Python, an interpreted programming language designed a while ago, became widespread because of its adoption by large organizations and because of its rapid growth due to contributions by a vibrant open-source community. Jewels of ideas from Matlab and R were incorporated in Python, and a large collection of Python libraries were developed for myriad tasks. AI tools moved to Python, and it is now fairly standard to use AI libraries in Python on a Linux machine. Jupyter notebooks for building prototypes in Python have become popular as they offer an integrated development environment and can access remote machines.

13.1.3 AI Frameworks

One of the earliest frameworks for ML was Torch (2002), based on an underlying C library interfaced with the LuaJIT scripting language. For research work in neural networks, one of the pioneering frameworks developed at the University of Montreal was Theano (2007). Since differentiation provides the basis for gradient descent, Theano had a symbolic way to compute derivatives:

```
>>> import numpy
>>> import theano
>>> import theano.tensor as T
>>> x = T.dscalar('x')
>>> y = 4 * x ** 2
>>> gy_wrt_x = T.grad(y, x)
>>> f = theano.function([x], gy_wrt_x)
>>> f(3)
```

The output is value 24.

In the second decade of the twenty-first century, a large number of AI frameworks have been implemented due to the explosive growth of AI. One of the first ones was Caffe from UC Berkeley. It was written in C++ with a Python interface, used configuration files, and implemented CNN. It is fair to say that Caffe made a big impact in popularizing AI in academia and in industry around the world. TensorFlow, CNTK, MxNet, and others were developed soon after Caffe.

TensorFlow was rapidly adopted by industry, with lightweight libraries such as Slim built on top of it. You could write a complex architecture as a Python script. The flow of computation through the script declares a computational graph, which is executed in a session. Blobs of data, also known as tensors, get computed as the computation flows through the graph.

A simple example, which will output the value 31.0 for the derivative of $a + 5a^2$, is as follows:

```
>>> import tensorflow as tf
>>> a = tf.constant(3.)
>>> b = 5*a**2
>>> g = tf.gradients(a + b, [a])
>>> sess = tf.Session()
>>> sess.run(g)
>>> sess.close()
```

Keras makes life very easy for AI engineers; it is a high-level API for TensorFlow, and it can use Theano, CNTK, or MxNet as backend in addition to TensorFlow. Keras takes care of most common use cases and implements best practices for AI research. Here is an example script meant to illustrate the syntax of the XOR classification problem:

```python
from keras.models import Sequential
from keras.layers import Dense
import numpy as np

# XOR Quadrant problem
Q00 = np.random.randn(100,2) * 0.25
Q01 = np.random.randn(100,2) * 0.25 + np.array([0,1])
Q10 = np.random.randn(100,2) * 0.25 + np.array([1,0])
Q11 = np.random.randn(100,2) * 0.25 + np.array([1,1])
X = np.concatenate((Q00,Q01,Q10,Q11))
y = np.concatenate((np.zeros(100),np.ones(100),
                    np.ones(100),np.zeros(100)))

# define feed forward network to solve XOR Quadrant
    problem
model = Sequential()
model.add(Dense(10, input_dim=2, activation='relu'))
model.add(Dense(10, activation='relu'))
model.add(Dense(1, activation='sigmoid'))

# compile the keras model
model.compile(loss='binary_crossentropy', optimizer=
    'adam',
                metrics=['accuracy'])

# train the keras model on the training data
model.fit(X, y, epochs=100, batch_size=4)

# evaluate the trained network
_, accuracy = model.evaluate(X, y)
print('Train Accuracy =  %.2f' % (accuracy*100))
X_test1 = np.array([[.25,.25], [.25, .75], [.75,.25],
                    [.75, .75]])
X_test2 = np.array([[0,0], [0,1], [1,0], [1,1]])
y_test = np.array([0, 1, 1, 0])
_, accuracy = model.evaluate(X_test1, y_test)
print('Test 1 Accuracy =  %.2f' % (accuracy*100))
_, accuracy = model.evaluate(X_test2, y_test)
print('Test 2 Accuracy =  %.2f' % (accuracy*100))
```

A strong contender to Keras is PyTorch, which is gaining popularity in academia as well as industry. Unlike TensorFlow, where one has two parts, namely, declarative and imperative, in PyTorch it is a define-by-run framework. The gradient descent is defined by how your code is executed at the run-time, and every single iteration can be different. This makes it easy to get started with PyTorch and to debug models. One can place a Python breakpoint anywhere inside PyTorch code, which could very well be inside the routines implementing the forward and backward passes. TensorFlow's eager execution is inspired by this define-by-run imperative programming environment. It evaluates operations immediately to concrete values without going through the stage of building computational graphs to be executed later.

---

* Exercise 29   Dynamic Graphs

State the advantages of dynamic computation graphs over static computation graphs.

AI software frameworks have contributed enormously to rapid prototyping of AI solutions. Using very compact Python scripts, AI engineers can build complex architectures.

13.2   Let's Crunch Numbers

We know about Central Processing Units (CPUs), the brains in our laptops. A CPU has a few cores, each of which can perform fast sequential processing. To speed up the computation, Graphics Processing Units (GPUs) have become the workhorse of AI. Video game players have used GPUs for a long time, and due to the parallelism they offer, they can speed up AI computations significantly. A GPU has thousands of cores, each of which is much simpler and slower compared with a CPU core, which by working together can offer massive parallelism. Tensor Processing Units (TPUs) are specifically designed for tensor operations in deep learning and work nicely for CNNs. Application-Specific Integrated Circuits (ASICs) are dedicated chips in which we implement commonly occurring operations in hardware; they offer the highest speed but with no flexibility to change the design parameters because operations get hardwired in the silicon. Field-Programmable Gate Arrays (FPGAs) provide a middle ground between the high speed of ASICs and the flexibility of software running on CPUs/GPUs. The cost of hardware per floating-point operation (FLOP) has been falling steadily, which will facilitate further progress in AI.

13.2.1 Computing Hardware

In AI, numbers have to be calculated in two stages:

1. During training one has to perform a very large number of operations on very large datasets.
2. At inference time one has to perform the forward pass for every test sample in the real world.

In order to make the right decisions for choosing AI machines, one has to ask the following fundamental questions:

1. What hardware will be needed? Will it be an embedded processor, CPU, Graphics Processing Unit (GPU), TPU, or FPGA? Will it be a multi-GPU machine? Will it be multiple machines?
2. Where will the hardware be located? Will it be at the edge, the cloud, premises, or a colocation center?

For training, you will need significant computing resources. SGD involves minibatches of data getting pushed through a large DNN. The size of the dataset, the memory requirements for a minibatch, the size of the neural network, the expected number of training epochs, and the overall dollar budget will guide you in selecting the training hardware. At inference time, you need to run the forward pass, and the particular use case will determine your choices. See [224] for a comparison of different computing choices.

** Exercise 30   Memory Requirement

Suppose an AI model has N neurons and M trainable parameters. For a minibatch of B samples, compute the memory requirements for training and for inference.

13.2.2 GPU Machines

For the training hardware, GPU instances in the cloud are one option, but a long-term analysis may show that you are better-off financially by purchasing your own GPU machines. If you decide to use physical GPU machines, then you will need to install software tools and AI frameworks. This option will offer greater control because it is just like a conventional desktop- or server-based experience. If the machine comes with a desktop environment, then GUI-based tools will be available to you. Due to the high demands of AI machines, a number of vendors are now offering GPU machines with popular AI frameworks preinstalled. From a toy GPU laptop to a GPU supercomputer, many choices exist, and you can hit the ground running.

13.2.3 Cloud GPU Instances

This section shows you the new way of doing computing by giving you a glimpse of the work of an AI engineer, who is setting up a project in the cloud. There is a universe of tools available in the cloud, which facilitate AI and classical ML. In order to get started, you will execute the following steps:

1. To begin with, you will sign up with a cloud computing provider, e.g., AWS, GCP, and Azure. Let us say you choose GCP [230]. Your first step will be to sign up on the GCP homepage and configure your account.
2. For training AI models, set up a virtual machine (VM) image optimized for AI and ML that will have all the necessary frameworks and tools preinstalled for you.
3. Customize the VM image by selecting the number and the type of GPUs. This choice will determine your monthly cost; therefore, be prepared to see the cost jump up if you opt for multiple powerful GPUs.
4. Once the VM compute instance is running, run your training scripts on the instance. Remotely login (SSH) from your laptop into the VM instance, and copy files using *gcloud*, the GCP command line tool. Run *gsutil* commands to interact with the cloud storage where your data may be located. A storage bucket can be mounted on your VM instance.
5. To run the Jupyter notebook, set up networking on the VM instance to allow HTTP traffic.
6. Run the Jupyter notebook in the VM instance, and use the browser on the local machine to connect to it remotely using the IP address of the instance and the notebook's port. Debug your training sessions using visualization tools such as TensorBoard, where you can display losses, hyperparameters, predictions, and any useful information.
7. Use Linux commands such as *screen* or *tmux* to ensure the training epochs are not terminated when the remote SSH session or the local machine is shut down.

With increasing connectivity the experience of using physical machines and that of using virtual machines are becoming similar. Furthermore, with continuing advances in the cloud-based AI platforms, the end user can completely work in a browser-based environment, which provides managed Jupyter AI notebooks, shell terminals, and integration of various tools, without a need of running any remote commands on the local machine. One can even use a tablet as the local machine to build and deploy an AI model on the cloud. Therefore, the primary consideration guiding the choice between the two options is financial.

13.2.4 Training Script

At the heart of training of AI models is the training script. During the experimentation phase, the front end will be typically a flexible Jupyter notebook with supporting Python files. To provide an idea about the training code, here is a possible breakdown of the script:

1. You will define a dataset along with data augmentation transformations. This dataset will be consumed by a data loader, which facilitates shuffling of data and creation of minibatches. The dataset could be based on map-style data in which indices or keys are mapped to data samples. Alternatively, it could be an iterable-style dataset representing a stream of data such as one being read from a remote database or even logs generated in real time. If the dataset is small, then it can directly fit into memory. If there is enough memory, deterministic transformations in data augmentation can be precomputed and cached in memory. For larger datasets, precomputed deterministic transformations can be cached on disk to speed up the training session, and the remaining random transformations can be computed on the fly.

2. You will define your AI model class with its forward pass. A model will be instantiated from this class. The backpropagation algorithm will be called within a nested loop of epochs and minibatches. The data loader will be called repeatedly to produce minibatches. Note that the backward pass will be automatically performed by the AI software framework. A careful optimization of this nested loop is necessary because operations are done over many iterations.

3. You will display crucial performance metrics and predictions while the training session is under progress. Both training and validation metrics are tracked. Therefore, you will need two data loaders, one for training and another for validation. At the end of the training session, a final display is made to assess the performance.

13.2.5 Deployment

Training of AI models is part of the story, and the other part is inference. For inference, your particular use case is the primary consideration. How will the trained AI model be used in the field? Are there latency requirements? Is there a need to keep customer data confidential within a perimeter? What are the constraints on the financial cost? What are the power requirements?

If the computation budget is tight for the inference, smaller AI models such as MobileNet can be trained. To deploy larger models, by using post-training model optimization, the precision of the parameters is reduced, say from 32 bits to 8 bits, leading to a reduction in the run-time memory requirements. A second option is in-training model optimization, which prunes away weights that are close to zero, resulting in a reduction of disk storage for models. Furthermore, by adding support for operations on sparse matrices in the forward pass, weight pruning leads to a reduction in the run-time memory requirements. Research is being done on a technique called knowledge or model distillation, in which a large network is trained and then its predictive powers are transferred to a smaller model [92]. This could emerge as another way to compress AI models.

A large software engineering project has to be undertaken in order to deploy the AI model. Companies have been extending their offerings of tools to assist AI engineers in the process, for example, TensorFlow Extended allows deployment on a server, a cloud instance, or a device. Ideally, you should deploy the trained AI model in the same programming language which was used for prototyping. Since Python is widely used as a prototyping language, it should be used in the production environment unless there is a critical reason to go for a faster language such as C/C++.

There are many variables involved in comparing hardware choices. To get a concrete feel for how a CPU compares with a GPU, one can expect anywhere from 5x to 100x speedup by using the latter. Having a GPU library, such as cuDNN [229], which has been optimized for deep learning, leads to a 2-3x speedup over an unoptimized library. This is true for CPU as well, and use of CPU-optimized deep learning libraries can provide a 2-3x speedup for inference, for example, see [236]. Due to the importance of benchmark suites for different use cases, a new initiative called MLPerf has been launched [234].

13.3 Speeding Up Training

13.3.1 Data Parallelism

To understand how GPUs fit with AI, note that a GPU consists of thousands of cores, which can run in parallel. A software program is written in the form of grids, blocks, and threads, which map to the GPU hardware. Matrix multiplication is a core component of AI computation and can be massively parallelized on a GPU. The backpropagation computation is off-loaded to GPUs. Say there are two GPUs, A and B. To keep both busy, the gradient computation is done in parallel,

$$\Delta_A = A(M_{t-1}, D_A),$$

$$\Delta_B = B(M_{t-1}, D_B),$$

where M_t is the AI model at time t and the minibatch has been divided into D_A and D_B. The updates are sent to the CPU, which aggregates the gradients and updates the model,

$$M_t = M_{t-1} + \Delta_A + \Delta_B,$$

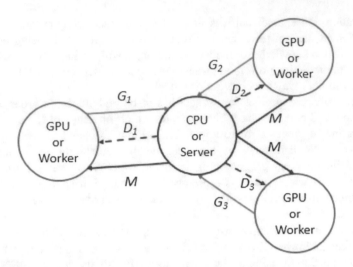

Fig. 13.1 SGD-based training can be parallelized in multiple ways depending on the model size, the training data size, available devices, available memory, and communication overhead. In the figure, we show one possible way to achieve data parallelism. Data is read from files, preprocessed, split into shards, and sent to devices. Each device computes gradients (G_i's) of the loss function with respect to the model parameters on its slice of data (D_i's) and sends them to a device, which adds them and updates the model M. The updated model is sent to the devices, and the whole process is repeated. All devices are kept busy by pre-fetching data chunks and preparing it for the next stage in a pipeline manner while devices are still working on the previous data chunks. The updates can be synchronous or asynchronous. The gradients can be compressed

and sends the updated model to the GPUs. See Fig. 13.1. This is *synchronous* data parallelism. One should ensure that GPUs receive data without any bottleneck. Smaller training datasets can fit in the GPU or CPU RAM memory. If a training dataset cannot fit into these memories, then high-speed external memory such as a solid-state drive is used. By carefully establishing a processing pipeline with pre-fetching of data and ensuring that no device is idle, one can eliminate or reduce communication bottlenecks.

In *asynchronous* data parallelism, suitable for the case when training is distributed over several worker machines, the model at the master parameter server (analogous to the CPU) gets updated whenever it receives an update from a worker; an update is then sent to other workers. This causes the stale gradient problem as the gradients are still being computed on some workers on a stale model. Therefore, asynchronous mode should be used after careful consideration of its impact on the test accuracy. If the drop in the test accuracy is high, then synchronous multi-machine training is used. In synchronous multi-machine training, some worker machines may fail, get stuck, or become too slow. The problem of straggler workers is solved by using more workers than needed and updating the model as soon as the first N workers send their updates.

For parallelization of reinforcement learning, one natural way is to parallelize different rollouts. Synchronous parallelism may become slower when these simulation trajectories are of varying lengths, though it has better training stability. For faster training, asynchronous approach should be tried, which achieves higher throughput. Care should be taken to ensure that asynchronous methods do not suffer from training instability issues and lower performance due to stale policies.

13.3.2 Delayed and Compressed SGD

There is a significant communication overhead between the Central Processing Unit (CPU) and the GPUs in the multi-GPU case and between the parameter server and the workers in the distributed case. Therefore, it is worthwhile to try approaches which compress the gradient values. Since a significant number of parameters have small gradient updates, there is overall a lot of redundancy in their values, leading to high compression, and this reduces the communication overhead. Gradients with high magnitude are sent, and this relates to the topic of delayed and compressed SGD. For certain applications, there is no significant degradation in the test accuracy, and empirical results suggest that SGD is robust to compressed and/or delayed stochastic gradient updates. See a recent reference [134] for more details.

> *A number of technical variables have to be carefully considered to determine the right hardware platform to train and to deploy AI models. Financial costs and use cases play significant roles in constraining the choices available to AI engineers.*

13.4 Open Ecosystem and Efficient Hardware

One of the strengths of AI research has been open-access, public-domain, and community-wide effort. Code is shared on Github [231], and papers are shared on ArXiv [227, 228]. Pretrained AI models and training datasets are downloaded for research. This is praiseworthy because it has allowed critical mass to build in AI. Furthermore, on AI software websites [233, 237, 238], there are extensively documented examples.

In the future, we will witness development of new innovative tools and further development of AI solutions will become easier, flexible and impactful. Highly complex and gigantic AI models pushing the state of the art to new heights will gain a foothold as part of this continuing progress.

At the same time, we will see novel *energy efficient* hardware being used for compute-intensive tasks. The carbon footprints of the training of large AI models rival the lifetime energy consumption of several cars, which is harmful to the environment. One of the reasons is the separation between the processing units and the memory, and data needs to be constantly transferred back and forth between them. Research is underway which allows in-memory processing using just 1% of the energy used when the same network is trained with conventional methods; see [106]. We can expect such breakthroughs in hardware, which will speed up computation with minimal use of electrical power.

Finally, it remains to be seen whether analog, neuromorphic, and quantum computation paradigms will be part of these advances. *Quantum computation* has attracted attention in recent years because qubits (quantum bits) can be in superposition of 0 and 1 and therefore offer parallelism. However, when measurement is made, then a qubit collapses to a classical bit, which places constraints on what can be achieved overall since eventually one has to read out the correct answer. In quantum mechanical language, wrong computation paths have to cancel out with destructive interference, and only the paths leading to the correct answer should reinforce each other with constructive interference. AI software frameworks are being expanded to include quantum computation which can be simulated. For example, TensorFlow Quantum is such an extension.

13.5   Chapter Summary

AI has ushered in a new way of programming supported by open-access AI software frameworks. The history of computing has witnessed several trends such as desktop computing, C/C++ programming, scientific programming, proprietary commercial software, open-access software, and cloud computing. Today the norm is prototyping in Python in the Jupyter notebook's IDE with the help of a large selection of libraries. AI software is becoming easier to use and to deploy. TensorFlow, Keras, PyTorch, CNTK, and MxNet are examples of AI software frameworks. GPU and specialized chips are expediting the training of AI models. Cloud computing has emerged as an alternative to physical machines for AI work. Combined with data parallelism and with compressed and delayed SGD, AI engineers can now expect to train their large models on large datasets in a matter of a couple of days or even hours. Training and deployment of AI models has to be carefully engineered and requires knowledge of the wide variety of options which are available. In the future, we will continue to see advances in both software and hardware, which will facilitate adoption of AI in diverse settings.

Chapter 14
AI and Business

> The first rule of any technology used in a business is that automation applied to an efficient operation will magnify the efficiency. The second is that automation applied to an inefficient operation will magnify the inefficiency.
>
> Bill Gates

> If you're trying to create a company, it's like baking a cake. You have to have all the ingredients in the right proportion.
>
> Elon Musk

AI solves problems and improves the state of the art. It does so within an organizational context. Goals may include improvement of existing products and services or creation of altogether new ones. It may be just streamlining of current business operations by removing inefficiencies. Organizations can be in either the public or private sector, and the same principles of effective use of AI will apply to both.

In this chapter, we outline guidelines which organizations should follow so that they are successful in this regard.

14.1 Strategy

In concrete terms, the strategy of the organization will start with a drawing on white board with two halves. On one side is the present state of your products, services, or business operations. On the other side is the same but with AI included. This is your long-term strategic vision.

Strategy is the set of tools and transformations which you need to employ to translate that vision into reality. This will include bringing in new talent and expertise, educating executives and employees about AI, expanding software tools and hardware infrastructure, and deploying tools to measure progress.

14.2   Organization

It is the people who will build and execute your strategic vision. And people need the right organizations to function effectively and productively.

Despite progress in software tools to build AI solutions, AI remains a difficult exercise from a technical point of view. It is different from building traditional software. AI demands expertise in mathematics, computer science, and software engineering. Some AI models can be fairly big with complicated loss functions, and, furthermore, they can be just one component of a larger system.

The first thing which is a key to success is to bring in people who are capable of building a solution, irrespective of whether it is a simple one or a complex one. It is competitive to hire such engineers and scientists. By understanding what incentives motivate such people to build their careers in your organization, you can get an edge over other businesses.

The AI team should be embedded within the company in such a way that it provides them autonomy to experiment with ideas while being connected with the strategic vision and the rest of the organization. See Fig. 14.1. Different members of the AI team will work on shared AI models and even shared layers of the same model, and this brings in challenges regarding coordination of such work. These team-level challenges should be properly understood during the formation of the team and the execution of AI projects.

14.3   Execution

AI starts with the real-world data. The very first thing which must be done is to expose the AI team to the actual data and the desired output. If the data does not yet exist, it should be collected early on so that one is able to eyeball it in order to develop insights into the complexity of the problem.

The AI team should answer the following two questions:

1. If the problem has to be solved by classical ML, statistics, and data science, then what will the solution look like? Perhaps you already have such a solution in place. A related question is whether it can be improved further using the same techniques.

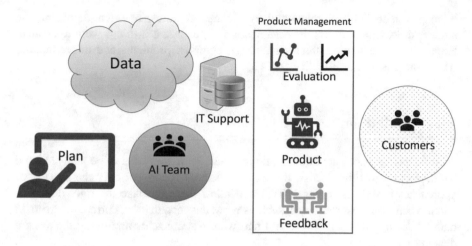

Fig. 14.1 To have a successful business in AI, emphasis should be placed on the data and constant evaluation of the results. The AI team is a highly specialized team, which should be integrated well in the organization

2. If the problem has to be solved by AI, then what will you be gaining or losing? What will be the risks and the rewards? What are the technology readiness levels of the proposed and existing AI solutions?

For large-scale projects, teams which are responsible for data acquisition and annotation should be built. Work can also be done by contracting it out to companies which crowdsource the tasks. AI involves much more than running the SGD algorithm on a deep neural network. The entire data science pipeline has to be implemented and maintained. Such a pipeline typically has the following components:

1. Data collection
2. Data cleaning
3. Data munging
4. Exploratory data analysis
5. Data annotation
6. Setting up the software environment
7. Setting up the hardware environment
8. Experiments
9. Research
10. Debugging of the AI models
11. Large-scale experiments
12. Deployment of the best AI model

Not all AI solutions have to be built in-house. The right decisions have to be made, and quick experiments have to be done to assist in this decision-making. Early feedback from customers, internal or external, has to be incorporated continuously.

In a rush to build products and to add new features, research should not be neglected. Research allows the organization to adapt fast if things were to change. Research should involve thinking through complex problems and understanding what solutions will be effective and why.

14.4   Evaluation

AI products need constant debugging and testing. Errors have to be analyzed and fixed. As part of this process, the AI team should report not only the success of approaches but all the failures. This will allow them to make the right decisions about where to invest resources, such as more data acquisition, correcting errors in data labels, upgrading software and hardware resources, or improving the system design.

Accuracy evaluation metrics such as ROC and PR curves should be presented to management regularly, along with the training time and the inference time. Decision-makers should educate themselves about these terms. In productization of AI, the inference time plays a big role in determining whether customers will see value in using AI. One has to compare different scenarios to make a good system design. How will the models be implemented in the cloud versus the edge? What is the use case for the cloud? Will it be running the AI model off-line or online? At the edge, will there be a single stream of data (e.g., for a mobile application) or multiple streams of data (e.g., for driving assistance in cars)? AI models have to be accurate. In real-world applications, they have to be fast and be able to meet latency and throughput requirements. Both the accuracy numbers and the inference times should be tracked for continuing improvements and to guide the best trade-off between the two.

Collaboration rather than competition should be encouraged between team members and between teams. No one person can achieve everything, especially when it comes to designing a complex system. Only teamwork which improves over time through tight integration with the rest of the organization and constant evaluation of results is the answer. Since AI is a fast-evolving field, integration with the wider community is helpful. This can take the form of writing papers and giving talks at conferences.

A number of organizations have been investing in AI, but a significant fraction of them report difficulty in scaling the models to their needs and in successfully deploying these models. Therefore, it is imperative that the design, execution, and evaluation phases are implemented correctly because of the specialized nature of such projects.

> ### * Exercise 31   AI Organization
>
> Read an article on the topic of AI written from a business perspective, which discusses the economic impact of AI in the coming years and whether organizations are being successful in adopting AI.

14.5 Startups

The above discussion applies to startups equally well. Venture capitalists and founders should start with the development of a strategy and build organizations from the ground up, which support the particular AI strategy. They have a greater chance of success because they do not have to face internal resistance due to the established ways of doing things.

It could happen that the problem can be solved by classical ML and data science. Such a classical solution should result in a baseline product. Most likely it will not be good enough to differentiate itself from competitors, and there will not be much barrier to others reproducing. Exercise in AI will bring a huge discipline to your overall effort by its emphasis on the data and on the evaluation of results, and therefore, it will allow you to outperform the baseline.

At the same time, in 2020 as these words are being written, AI startups are facing tough competition from big technology companies, which have larger datasets and bigger computing budgets. Startups often have to rent out cloud computing resources from the same big technology companies. Therefore, it is imperative for startups to develop a strategy, which will allow them to compete against big tech effectively. By addressing niche market needs and by creating valuable intellectual property, they will have a greater chance of success.

> ### *** Exercise 32   AI Startup
>
> Write a business proposal seeking funds to launch your startup in AI.

> *AI is a disruptive technology, and business leaders should make investments in AI with a sound understanding of the strategy, organization, execution, and evaluation needed to support the effort.*

14.6   Chapter Summary

AI has a transformative potential to make business operations more efficient, add value to existing products and services, and to create new products and services. Business leaders should keep their organizations at the cutting edge of innovation by strategically evaluating the impact of AI on their business and on their target market. By giving careful thought to the long-term strategy and by building the right organization to execute it, investment in AI will lead to long-term rewards. AI requires hiring of a highly specialized team, an emphasis on the data, and building of an infrastructure to constantly evaluate and improve the results. It should be kept in mind that AI is much more than just algorithms and mathematics. A huge, disciplined effort must be spent on implementing and improving the underlying data engineering pipelines. By bringing different components together, a business will stay competitive in a rapidly changing technology landscape. See [99, 242] to explore AI from the business perspective.

Part III
The Road Ahead

The future rewards those who press on.
I don't have time to feel sorry for
myself. I don't have time to complain.
I'm going to press on.

Barack Obama

Chapter 15
Keep Marching On

When I look into the future, it's so
bright it burns my eyes.

Oprah Winfrey

I have always been convinced that the
only way to get artificial intelligence to
work is to do the computation in a way
similar to the human brain. That is the
goal I have been pursuing. We are
making progress, though we still have
lots to learn about how the brain
actually works.

Geoffrey Hinton

In Chaps. 4 and 5, we learned the key reasons for the success of AI, which is its phenomenal expressive power, SGD not getting trapped in local minima, and ingeniously designed AI solutions trained with adequate data. The previous chapters have showed the brilliant work which has gone into the design of customized AI models which are amenable to the problems at hand so that their carving capacity is utilized efficiently. All these reasons have come together to make AI successful.

Despite the impressive work which has already happened, one thing which can be said with certainty is that most of the progress in AI really belongs to the future. We have definitely had a great start early this century, and it paves the way for further progress. The state of the art does suffer from shortcomings, which have to be overcome. For those who want to go where no one has gone before, AI offers a wonderful opportunity, and the frontier ahead is unlimited. Let us look ahead.

© The Author(s), under exclusive license to Springer Nature Switzerland AG 2021   269
S. Dube, *An Intuitive Exploration of Artificial Intelligence*,
https://doi.org/10.1007/978-3-030-68624-6_15

15.1 Robust AI

15.1.1 Adversarial Examples

Current AI models suffer from an embarrassing problem of adversarial examples. What can be demonstrated is that we can make an AI model misclassify a positive sample as a negative. Given a cat classifier $F(x)$, it can be nudged towards making a mistake by adding a carefully computed, visually imperceptible noise δ to any positively classified cat image x such that

$$F(x) = \text{Cat},$$

$$F(x + \delta) = \text{Not Cat}.$$

How is this possible when AI has been winning the ImageNet competitions? Even a human baby won't be fooled by that. Moreover, we can go in the opposite direction and find obviously negative samples z for which

$$F(z) = \text{Cat}.$$

These are strange-looking patterns or plain noise, and therefore no human will ever make such obvious blunders. A conclusion is that AI doesn't know what a cat is, and it is not robust because it can be tricked into making mistakes. See Fig. 15.1 for a perturbation adversarial example. See Fig. 15.2 for a noise-like adversarial example. For seminal work on adversarial examples, see [151, 210]. Adversarial examples make us wonder whether current AI systems have the required level of security, that is, whether they can be attacked by an adversary who wishes to bypass or break the system. An example is a vandalized stop sign with a little adversarial sticker, which can fool a self-driving car into misclassifying it as something else.

Recall that the output is a function of the input and trainable parameters,

Fig. 15.1 Perturbation-based adversarial example. AI returns the correct answer "Taj Mahal" on the left image. It can be fooled to return "Not Taj Mahal" for the right image by adding a visually imperceptible perturbation to pixels. This suggests that it is easy to step outside the carving anywhere, with most of the volume being on the surface of a twisting and turning manifold in a high-dimensional space. (Courtesy the Metropolitan Museum of Art, New York)

Fig. 15.2 For the second kind of adversarial examples, a properly trained AI model with high accuracy on Taj Mahal images can be fooled to return "Taj Mahal" on a noise-like image. This suggests that carving cuts are extending into the complement of the manifold, possibly because of lack of enough negative training data

$$y = F(x, \theta).$$

In backpropagation, x is fixed and the derivative is taken with respect to θ. For adversarial examples, θ is fixed and the derivative is taken with respect to x,

$$\frac{\partial F}{\partial x}.$$

For the first kind of adversarial images, instead of minimizing the loss function, we maximize it. The input x is modified under the constraint that the norm of δ is within a threshold. An easy way to ensure the latter is to take the sign of the gradient and multiply it by a small parameter α.

For the second kind of adversarial examples, one takes any initial image z' and performs many gradient descent steps in the input space to minimize the loss function as if the ground truth is a cat in order to make it more cat-like. If you choose z' to be a cat image, you will presume it will become more cat-like. On the contrary, it gradually becomes distorted and noisy and departs from being a cat. Despite that, due to the gradient descent, the AI model keeps on increasing the probability of the image being cat. The root reason for both is that despite its huge expressive power compared to classical ML, in terms of the number of linear regions, AI makes mistakes in carving out manifolds and in approximating functions.

The cat manifold is highly complex and is bumpy, snake-like, and irregular. If the surface of the manifold has turns and twists at very fine scales, then AI will suffer from the first type of adversarial examples if carving is unable to capture all the fine details. You can always step across to the other side of a carving cut and make mistakes. In high-dimensional spaces, most of the volume of a manifold is on its surface, and therefore it is easy to perturb a positively classified sample to push it out of the carved manifold; see [50] for this hypothesized explanation. As images become larger, it becomes easier to perturb them with smaller perturbations. The norm of the required perturbation falls off as $O(1/\sqrt{n})$ where n is the input resolution and therefore shrinks to 0 as $n \to \infty$; see [50]. In Fig. 1.3, almost every image is at the boundary of the carved manifold. For the second kind of adversarial examples, note that the manifold is very high-dimensional, and therefore any practical negative dataset will be sparse due to the curse of dimensionality, making AI unable to carve the manifold out from all sides. There are just too many sides of an object in a high-dimensional space. In Fig. 1.3, if negative data is sparse, a convex polytope may not be split into cats and not-cats, where the latter include noise-like images. In other words, the cuts extend into the complement of the manifold, which means the negative points get included in the carved manifold; see [50] for this hypothesized explanation. The above hypothesized explanations are based on intuition backed by mathematical insights, and the exact causes for adversarial examples will be discovered with extensive experiments on actual datasets in the future. We can draw the conclusion that the hierarchical feature representation built by deep learning is not robust despite being superior to classical ML.

15.1.2   Learning from Human Vision

We can learn from human perception, which is robust to adversarial examples. Though it can be fooled by cleverly designed optical illusions, that's hardly embarrassing. On the contrary, illusions illustrate the human mind's extraordinary ability to fill in missing evidence in the confirmation of a hypothesis and to see things holistically in terms of relationships. Often an illusion is an image in which the human mind can generate two valid hypotheses.

Our perception uses generative models and it applies a fusion of multiple forms of evidence. It can build a holistic idea of the scene hierarchically in a robust manner. Motivated by gestalt principles, making use of part-whole hierarchy in inference tasks has been a theme in pattern recognition approaches [86, 147]. The manifold learning hypothesis states that parts and whole are manifolds embedded in a larger space and a part-whole hierarchy makes use of the relationships between these manifolds.

When we see an eye of a cat, we can predict the location, size, and attributes of its other eye. The presence of the second eye in the predicted location reinforces our belief that these are eyes and, in turn, the two eyes predict other parts of the cat. Different parts predict the attributes of the face, which, in turn, confirms or rejects the micro-evidence of parts. One can call it consensus building or voting. The idea of voting is powerful and has been used in Random Sample Consensus (RANSAC) and the Hough transform in computer vision. Mutual reinforcement of consistent evidence and attenuation of inconsistent evidence is belief propagation. We employ generative parameters of parts and of the whole in these feedback loops, and this makes our perception invariant to transformations and robust to perturbations. See Fig. 15.3. We have to carve the manifolds by exploiting their compositional topology; the whole manifold is constructed from the lower-dimensional manifolds of the parts in a compositional process.

Capsule Neural Networks (Capsnets) are inspired by the above observation and seem to be more resistant to adversarial examples. The idea is to have a data-driven method which learns the latent attributes of parts by capsules of neurons and builds

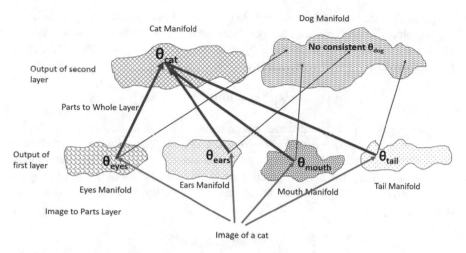

Fig. 15.3 An alternative way to implement AI models by fusing consistent micro-evidence. Visual recognition of parts and the whole in terms of their attributes, which include poses, denoted by Θ is done in such a way that all evidence which is consistent gets combined together to increase the probability of the final decision and spurious inconsistent evidence and unsubstantiated hypotheses are rejected

the part-whole consensus. See [89, 183] for details. Ideally, the learned attributes should have a semantic interpretation such as pose and orientation. Segmenting parts is challenging when it is done explicitly, and therefore a Capsule Neural Network (Capsnet) strives to perform it implicitly.

Here is the basic intuition. A capsule of neurons is a part detector which outputs a vector of k parameters (latent variables or attributes) describing a given part,

$$(\theta_1, \ldots, \theta_k) = f(X),$$

where X is a receptive field of the given image or of a feature blob produced by a few initial convolutional layers. There are p such part detectors. There is a learned relationship between outputs of capsules at different hierarchical levels,

$$(\phi_1, \ldots, \phi_r) = g(\theta_1, \ldots, \theta_k) = g(f(X)),$$

where g is a capsule detector of a whole object of a given class and it outputs r parameters of the whole object based on the output of a particular part detector. Thus, there are p capsules for the part-whole mapping, the part-whole capsule g_i for the part capsule f_i. For recognizing cats, the p parts are ears, eyes, tail, etc., and the whole object is a cat. The prediction for the whole object over a bigger receptive field Z is

$$u(Z) = \sum_{X \subseteq Z} \sum_{i=1}^{p} w_i(X) g_i(f_i(X)),$$

where $u(Z)$ is the r-dimensional predicted vector and the coupling weights initially are uniform. The weights are obtained from underlying log-likelihood weights b_i using the softmax function. Initially, $b_i = 0$ and therefore $w_i = 1/N$, where N is the total number of parts in the receptive field Z,

$$(w_1, \ldots, w_N) = \text{softmax}(b_1, \ldots, b_N).$$

In other words, to find the attributes of the whole object in an image region Z, find the attributes of its parts inside that region ($X \subseteq Z$), and then take the weighted sum of the attributes predicted by different parts (the weights are w_is), where the weights indicate how likely it is that the putative parts are indeed parts of the whole object. The log-likelihood weights are iteratively updated to determine their consensus with the whole object,

$$b_i = b_i + d(g_i(f_i(X)), u(Z)),$$

where d is a suitably chosen distance between two vectors, such as dot product. The dot product is the magnitude of the "consensus" between the whole and the part. The prediction vector $u(Z)$ is updated using new weights. This process is repeated several times to establish mutual consistency. The weights signify the agreement of the parts with the whole. These weights are iteratively refined by pruning away inconsistent evidence and strengthening consistent evidence. See Fig. 15.3, which illustrates the approach for the cats versus dogs problem.

This will be repeated for C different classes using the same part capsules but different object capsules. If for another class there is no consensus, then the belief in that class will go down. The whole network can be trained using backpropagation under the classification loss. To ensure that the attributes are meaningful latent variables, a generative branch with the reconstruction loss is added. Therefore, the technique is in between discriminative AI and generative AI. A practical implementation of Capsnet will approximate receptive fields by patches. One can anticipate continuing progress in this direction based on robust segmentation of parts, inference of semantically meaningful attributes of parts, and explicit Bayesian inference.

Bayesian inference plays a role in human perception [110] in merging evidence. See [119], which explores how Bayesian inference can be used to learn handwritten characters from the world's alphabets with only a few given examples. Probabilistic graphical models (PGMs) have been used in Bayesian statistics to model mutual interdependence of random variables; they come in two flavors, namely, Bayesian networks and Markov random fields. Therefore, in the future, we can anticipate PGMs being used on top of the outputs of CNNs. In the near future, it will be interesting to comparatively investigate and visualize how different approaches which are based on CNNs, capsule networks, and Bayesian inference carve out the manifolds and their respective optimization landscapes.

Human perception is complex, and a reasonable approach to model it will have several components; CNNs, Probabilistic Graphical Model (PGM)s, Bayesian inference, generative models, robust part detection, part-whole synthesis, holistic scene segmentation, and a world model will all come together.

Note that generative AI can solve the problem of perception if we can figure out how to transform the latent parameters of an image in order to produce a target image. To recognize a cat, just check whether you can generate it. In the future, we will see advances in generative AI. Part-whole synthesis will be applied to generative AI models. A scene can be generated by first generating its parts and then putting them all together in a whole. A future robot will look at a scene of wind gently moving the leaves of a tree. Using self-supervised learning on this image sequence and by applying the coherency of motion, it will learn how to accurately segment individual leaves and use these segmented leaves as the ground truth to train a leaf generator. The generative parameters of a leaf can then be changed to spatially transform one single leaf into hundreds of thousands of leaves of a tree.

There has been criticism of the current deep learning approaches (see [139]), which stems from the fact that state-of-the-art AI is still not robust and works well only within constrained domains. However, it is really a matter of continuing progress, which does require investment in research and in computing power, and we can anticipate that AI models will become more robust with time. We have just scratched the surface at this time. We should keep in mind that the human mind has been pre-trained for many tasks by at least 3.5 billion years of evolution. It is the same pre-training which also allows a spider to weave its web, a bee to encode the location of flowers in its dance, and a hummingbird to hover before a flower. With AI, we have just begun the journey.

We can ask if there are biologically plausible alternatives to the biologically implausible backpropagation-based learning. Does the human brain use pyramidal neurons which can do both forward and backward passes? Does it use predictive coding in which a layer of neurons predicts the output of the preceding layer and adjusts synaptic weights based on the prediction error? Is it dopamine-mediated reward-based learning that strengthens those connections which lead to accurate predictions? The answers could inspire new AI methods.

** Exercise 33   Bayesian Inference

Give an example to illustrate how Bayesian inference is used to update beliefs by fusing evidence. See Fig. 15.4.

15.1.3   Fusion of Evidence

We build further intuition about how fusion of evidence works. Consider a very simplified example. Suppose you are seeking to estimate price of a house. One

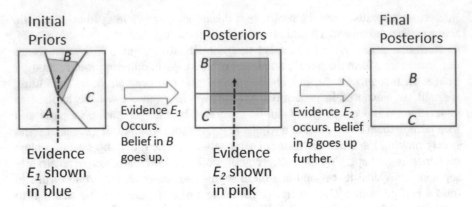

Fig. 15.4 Iterative Bayesian inference. On the left, priors for three hypotheses A, B, and C are shown in a universe of outcomes. Evidence E_1 is shown in the shaded light blue region. If E_1 occurs, posteriors for the hypotheses change. Note that belief in A goes to zero, whereas for B, it increases. These become priors for the next iteration. If further evidence E_2 which supports B occurs, then belief in B goes up further

appraiser estimates it to be a random variable X_1 with mean μ_1 and variance σ_1^2, and other appraiser estimates it to be X_2 with mean μ_2 and variance σ_2^2. How will you combine the two estimates? Let us linearly combine the two variables

$$Y = (1 - \alpha)X_1 + \alpha X_2.$$

It is a weighted combination. The weights should reflect our trust in the two estimates. The lower the variance, the higher the trust. An estimate with a lot of uncertainty as quantified by a high variance should be trusted less. We want the variance of Y to be minimized for the most certain estimate. Since we have

$$\sigma_Y^2 = (1 - \alpha)^2\sigma_1^2 + \alpha^2\sigma_2^2,$$

by differentiating the above with respect to α to minimize it, we find

$$\alpha = \frac{\sigma_1^2}{\sigma_1^2 + \sigma_2^2}.$$

We can rewrite the weighted combination as

$$Y = X_1 + \alpha(X_2 - X_1).$$

If X_1 and X_2 are Gaussian, then Y is Gaussian, and then its mean is the best estimate of the price of the house.

This is similar to the Kalman filter which is applied to combine evidence in various applications. Consider the application of the Kalman filter in the simultaneous localization and mapping algorithms for robotics in which there is a concept of a state. A robot has odometry evidence and landmark observation evidence. Odometry allows a navigating robot to predict based on the previous state and observing visual landmarks allows it to bring in another evidence. The two are combined using the Kalman gain K which plays the role of α above,

$$(1 - K)(\text{Odometry}) + K(\text{Landmark}).$$

To compute K, the scalar variances are replaced by covariance matrices, and the reciprocal is replaced by the matrix inverse. The Kalman filter provides the closed-form solutions for Gaussian case and linear models. For general non-linear, non-Gaussian case, one can employ particle filters, which use Monte Carlo approach to approximate the probability distributions and to combine evidence.

Bayesian inference subsumes the above approaches in a general framework. To see this beautiful result, first consider the continuous version of the Bayes inference rule. We now have continuous probability distributions. The goal is to find the mode of the posterior probability $p(x|y)$ where x is the state and y is the observation. To derive the Kalman filter, plug the Gaussian distributions in the Bayes inference rule. To obtain the maximum a posteriori (MAP) estimate of the state, take the logarithm of the two sides of the Bayes rule, and differentiate with respect to x to maximize. This yields the Kalman filter (see [32]), as a special case of Bayesian inference.

In human intelligence, we noted that most likely we fuse evidence. An eye provides an evidence of the face and of the other eye which is analogous to a robot using odometry to predict its next state. Other pieces of evidence of the face and of the other eye are analogous to the observation evidence. These all get combined in Bayesian framework which attenuates or strengthens beliefs. We can anticipate that in the future, an AI model will compute and then fuse all these evidence in a hierarchical, iterative, and dynamical manner.

15.1.4   Interpretable AI

One of the complaints against DNNs is that they are not interpretable. We want AI to explain why it classified something into category A and why not in category B. We have multiplications and additions going on in the hidden layers which yield feature blobs as the forward pass computation progresses from the input to the final answer. It is hard to interpret what exactly happened in the computation which led to the particular output.

Current methods which shed some light on the working of an AI model on a given input revolve around tweaking the input to see how the output changes accordingly. We run backpropagation from any output neuron or hidden neuron to the input and examine the gradients with respect to the input. For visualization, the model parameters θ are fixed, and the derivative of the output of the target neuron is taken with respect to the input x, and then by taking the magnitude of the gradients,

$$\left|\frac{\partial F}{\partial x}\right|,$$

one obtains the visualized gradient image. The gradient image tells us which pixels affect the output of the target neuron most. We obtain visualizations for neurons with high activation values to find out what input features activated them. A related approach is to mask out different regions of the input to see the impact on the activation of the neurons,

$$F(M(x), \theta),$$

where $M(x)$ is a modified input obtained by setting the pixels of a region to zero. If the output changes significantly, then the masked region played an important role in the decision-making process by the AI model. These masks can be generated randomly, and a weighted combination of these masked inputs where the weights come from the score of the target class yields an interpretable saliency map; see [161]. An alternative approach is to use a simpler interpretable ML model F' on randomly chosen samples in a local neighborhood of the given sample x in the input space to evaluate which features are important. Therefore, F' provides a surrogate, locally interpretable, model-agnostic explanation of feature importance; see [179]. For a more recent work, see [10], which shows how individual neurons of CNNs and GANs can be interpreted by looking at the interdependence between neuron activations and network inputs and outputs. Interestingly, they show that sometimes a single neuron can encode a human-interpretable visual concept. This supports the grandmother hypothesis which states that there should be a neuron which fires exactly when you see your grandmother.

Progress in interpretable AI will eventually mean building more robust AI which will be amenable to interpretation. If we are able to see how parts were detected in a hierarchical fashion and how the micro-evidence got fused to do so, then it will tell us at a high level how the model arrived at its decision. In the future, AI will be able to say, "Well, I had strong evidence for the facial features of the cat which indicated that the cat's face was turned 30 degrees towards the right and its feet indicated that it was sitting on the grass. Fusion of all the evidence made my belief in the presence of a cat go up to 99%. In the process, I was temporarily thrown off the track by a spurious detection of a tail which suggested a pose inconsistent with the one suggested by other parts. Segmentation of other objects in the scene confirmed that it was not a tail but a bunch of dry leaves of the same color as the cat. Therefore, I discarded that micro-evidence."

> *Convergence of several techniques such as discriminative AI, generative AI, and Bayesian inference, backed by an increase in computation power, will lead to robust and interpretable AI.*

15.2 AI Extraordinaire

Human intelligence is not based on doing gradient descent on tons of labeled data over training epochs. We use unsupervised learning. We just observe data. Unlabeled data doesn't mean that you can't learn from it. Labels may be implicitly derived. As a child observes food and spoon fall to the ground, much to the consternation of the parent, it gets labels on the current image from the next image in the time sequence of images which can be used to build a prediction model. The child learns that though most things fall to the ground, some don't, such as smoke or birds. Like in language modeling in NLU where one predicts the next word, this is world modeling in which we predict what will happen next. This is self-supervised learning. It enables us to predict a masked part of an image from its other parts and its one viewpoint from its other viewpoints. It enables us to transform one image to a related one. Its robust solution is likely to come from generative AI which learns latent generative representation for these tasks. This can be then followed by a supervised task-specific fine-tuning stage. In the future, AI will learn based on its own observations.

We have an inner sense of space and time. We have theory of mind, which allows us to put ourselves in the shoes of others. For us, a piece of text is not a sequence of words and a trajectory in an embedding space. It is a story in space and in time. One of the joys of reading is that we create a world in our mind when we read; see Fig. 15.5. The audiovisual representation and the textual representation of the same story will become convertible into each other. From a book, an AI model will be able to produce a movie and vice versa. A story has an external world of characters which evolves with time, and it has an inner world of feelings and thoughts of characters. A story is a dynamical system with states. We run temporal simulations of stories like a rollout in RL, except now we are working with high-level concepts and with people. This leads to counter-factual thinking and to common sense and reasoning. In the future, we can expect AI to do similar rollouts.

In the future, a self-driving car will be a self-learning car. It sees a yellow bus stopped ahead on the road. It reads text on the bus and concludes that it is a school bus. By observing children in the past, it knows that they can be unpredictable. By doing a temporal rollout based on the rules it has learned so far, it would assign a probability of children getting out of the bus and walking. See Fig. 15.6. The car is able to enhance its world model by learning new cause-and-effect relationships and logical rules. Statistical Relational Learning (SRL) seeks to learn logical rules in a data-driven manner. In the future, a combination of such techniques will make AI smarter with time. AI will reach great heights. One day, it may report a new logical rule it has learned,

$$(\text{Action} = \text{Good}) \rightarrow (\text{Mood} = \text{Happy}),$$

with coverage = 7% and accuracy = 96%. Though it notices good actions only 7% of the time among all its observations, it sees that they make you happy 96% of the

"Well then," the Cat went on, "you see a dog growls when it's angry, and wags its tail when it's pleased. Now *I* growl when I'm pleased, and wag my tail when I'm angry. Therefore I'm mad."

"*I* call it purring, not growling," said Alice.

"Call it what you like," said the Cat. "Do you

Fig. 15.5 A story is more than a text document. It is an experience in sights and sounds which unfolds in space-time. There is an inner world of characters. (Image credit: Lewis Carroll, the author, and John Tenniel, the illustrator; the first edition of Alice in Wonderland (1865), Macmillan & Co)

Fig. 15.6 A self-driving car will have to learn to do inference by building a world model. Interdependence between observations can be captured by Bayesian networks. Humans have a rich world model which allows them to understand cause-and-effect relationships

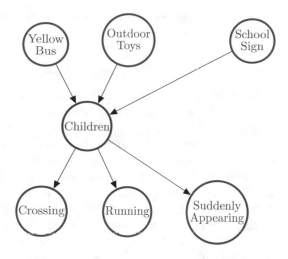

time. It is able to rediscover the truth seen by Buddha more than 2,500 years ago. This will be an example of AI extraordinaire. For future AI engineers and scientists, opportunities are just unlimited.

Dr. Seuss imagined fantastical and imaginary creatures in his books. A child needs to look at one illustration of a Schlopp, Snuv, or Guff in order to relate to it, recognize it, and draw it. In the epic poem of Ramayana, erudite and arrogant Ravana is shown to have ten heads. In Greek mythology, Pegasus can fly. These are examples which show that we can take familiar classes and tweak them in our imagination. They show that we have the capacity to work in a model space where points are whole manifolds. A model space is a parameterized map of manifolds in the same way the Mandelbrot set is a map of Julia sets. We have an amazing ability to transform one model into another in this space. Here is a simplified example. Suppose the model space M for real or imaginary beings is parameterized by values such as number of heads, number of tails, number of legs, etc. A human has one head, no tail, and no wings and is a point $x \in M$. A child moves to another point $y \in M$ for a human-like fictional character with one head, three tails, and two wings in a Dr. Seuss's book. For Pegasus, instead of no wings for a horse, the model is changed to two wings. For Ravana, instead of one head, it is increased to ten. In the same way, the child can tweak statistical hyperparameters, for example, he/she can imagine a world in which everyone has become Lilliputian or Gargantuan. He/she will transfer properties of one model to another model as prior beliefs and will update those beliefs with new evidence. We create new fantastic worlds in our imagination and transfer our familiar models to them. When you watch Toy Story, you don't need to learn everything about the world of toys from scratch because we transfer the attributes of our world to them. When we learn that toys don't move in the presence of humans, unless provoked, then we update that part of the world model. That's why we can learn from one example. It is like transfer learning in the model space. We can expect that AI of the future will learn in the same way, just with one example.

In other words, consider a function I which maps these generative hyperparameters $\theta \in \Theta$ to a manifold,

$$I : \Theta \to \mathbb{M},$$

where $\mathbb{M}$ is the set of manifolds. We can refer to Θ as the imagination space.

We noticed earlier that human intelligence is result of at least 3.5 billion years of evolution. The evolution of life has played a critical role in developing it to the present state. What if the same evolutionary forces are unleashed in AI? Robust part detectors which can detect and segment out parts and which can infer their attributes will lead to robust higher-level detectors. A frog can detect the motion of small objects well, which over time led the nascent human species to detect distant motion on the plains of Africa. More complex models get built on the top of simpler ones, and they constantly adapt through transfer learning. The best AI models will get used more frequently in an open community-wide ecosystem. AI combined with Artificial Life could set a process rolling forward which is similar to neural architecture search, a part of automated ML, but at a much bigger scale and over a longer time.

> *Relentless progress will lead to AI models at par with humans in their reasoning and learning ability. In fact, AI will develop extraordinary abilities over time.*

15.3   Chapter Summary

The future of AI is very bright. Though AI suffers from adversarial examples at the present time, new techniques which derive from the way human perception works will make it robust and interpretable. A combination of techniques will enable AI to visually analyze a scene in the same way we do. Part-whole hierarchy and holistic scene segmentation enabled by Bayesian inference built on the top of the outputs of CNNs will play a central role. The line between discriminative AI and generative AI will get blurred as both will capture the semantic essence of an object. For natural language understanding, AI will create an accurate depiction of what is happening in a piece of text by placing characters in space and time like we do when we read. Visual and textual versions of the same story will become two interconnected sides of one unified semantic representation. The hard boundaries between vision, NLU, speech, and synthetic content will disappear because they will represent different aspects of the same multi-modal reality. AI will constantly enhance its world model through a mixture of rollouts and observations and by learning cause-and-effect relationships. It will be able to recognize new real or imaginary things just with one example by working in a parameterized map where points are models, which can be mapped into each other, and attributes of one model can be transferred to another as priors. AI will be much more than deep learning in the future when it expands into all the above uncharted territories. Finally, AI combined with Artificial Life could initiate a large-scale experiment of AI improving itself over time.

Chapter 16
Benevolent AI for All

> AI research by itself will tend to lead to concentrations of power, money, and researchers ... this is not healthy. Even in a democracy, it's dangerous to have too much power concentrated in a few hands.
>
> Yoshua Bengio

> I think you can build things and the world gets better. But with AI especially, I am really optimistic. And I think people who are naysayers and try to drum up these doomsday scenarios – I just, I don't understand it. It's really negative and in some ways I actually think it is pretty irresponsible.
>
> Mark Zuckerberg

Human history teaches us that there are two sides of human nature. The Strange Case of Dr. Jekyll and Mr. Hyde is in fact not strange. "All human beings, as we meet them, are commingled out of good and evil," notes Robert Louis Stevenson, the perceptive author of the celebrated story. Individually we may believe that we are good, but collectively the two sides of our nature are very apparent. On the one side, we have public libraries and schools; on the other side, we have weapons of mass destruction.

Technology when put to good use is marvelous and propels the human civilization forward. In the wrong hands or due to plain shortsightedness, it can wreak havoc on humanity. Not only humans, it can endanger many other species on the Earth.

A long time ago, humans learned to make fire. In recent years, humans discovered nuclear energy. Both technologies have led to weapons. The same pattern holds for all forms of technology. For some reason, it seems that humanity is incapable of restricting the use of its brain power to good uses.

In this chapter, we look at the particular case of AI, no different from any other technology, and what can be done to ensure that it is used for the benefit of everyone.

16.1 Benefits of AI

The benefits of AI are immense. Wherever there are images, spoken words, written words, observations, and measurements, AI can make operations more efficient, create new products and services, and improve well-being. One doesn't have to restrict applications to big data and other conventional areas. Many diverse fields can benefit from AI.

Agriculture, medicine, research, education, engineering, the environment, exploration, governance, decision-making, social sciences, commerce, retail, communication, and transportation can all benefit. AI routinely powers human-machine interfaces, web search, recommendation systems, driver assistance systems, analytics, and commerce. It is being deployed in medicine, healthcare, biotechnology, laboratories, agriculture, transportation, research, and education. There is a growing awareness of the need to make AI human-centered and human-compatible.

AI is an extension of the human mind, our ultimate tool, and therefore it will be at the forefront. In the future, AI-enabled machines will be the first to land on the moons of Saturn and Jupiter and faraway extra-solar worlds. They will work alongside doctors saving lives and eradicating diseases. They will work on farms so that no human goes hungry. They will build houses so that no one is homeless. They will work with scientists making great discoveries. They will educate children and help them grow into fine adults. They will work with environment scientists to protect the diversity of life on the Earth.

AI has the potential of revolutionizing our economic system, bringing freedom to people, and eliminating human labor so that they can focus on intellectual, artistic, social, and scientific projects. Children born in the future will be able to take on grand challenges and save the Earth from pollution, degradation of the environment, and extinction of species, rather than worrying whether they will get jobs or not.

16.2 AI in Medicine

In particular, the future of medicine is in mathematics and computer science, with AI playing an increasingly important role. The expertise of physicians will get complemented and enhanced by AI tools; see [214]. Here are some examples.

In the book, we mentioned applications of AI for medical imaging [143, 153, 180, 235], for treatment planning [194, 215, 222], and for biomedical text mining [125]. There are many other potential applications. AI models have been developed for detection of diabetic retinopathy [71] and of arrhythmias [74]. Early detection of sepsis by AI can save many lives [188]. AI can be used for controlled delivery of insulin by an artificial pancreas based on reinforcement learning and for augmenting surgical methods with computer vision; see [76]. Data acquisition, diagnosis, treatment, and prevention will all benefit from developments in AI.

For pre-clinical research, AI can speed up the search of molecules which can inhibit enzymes involved in diseases [256]. See [189, 248] for a fine application of AI to the challenging problem of protein structure prediction directly from multiple sequence alignments data, which is critical for research in life sciences and medicine.

We anticipate that generative AI and discriminative AI will have an impact on medicine, in clinical and pre-clinical settings. Generative AI can be used to augment the training data and to create simulated videos. Discriminative AI will become a useful tool to analyze images, videos, sounds, and graphs. Combined with multi-modal healthcare data, it will provide a holistic picture of the patient's health and well-being and the most effective ways to enhance and maintain wellness. AI will make medicine more affordable and of higher quality.

16.3   Dangers of AI

The road to a better world in which we reap the benefits of AI is going to be a bumpy one. Dangers lurk everywhere which can easily derail progress.

What comes to your mind when you think of the future? Is it a Star Trek type of world in which we work together amicably and boldly explore the universe? Visualize a different scenario. In this alternate vision of the future, AI is under tight control of a few people who are living opulently while masses survive on the universal basic income, their jobs lost forever, and their boredom broken by addictive video games and meaningless activities provided by the state. Instead of a happy Star Trek future, it will be a dystopian world marked by disparity and oppression. By no means, an economy of abundance implies well-being of all citizens. AI has the potential to give total control to autocrats over their subordinated populations through invasive technologies which monitor every citizen, who will not have privacy and will be subject to 24/7 surveillance. Citizens will be watched and tracked continuously with money being made off their personalities, dreams, and needs in a surveillance capitalism. AI is capable of creating fake content which looks real and therefore can be misused to control people. Imagine a world in which you are arrested by a police state on the basis of a fake evidence.

AI depends on data, and there is a human footprint in it, from the time of its creation to the time it is used by AI scientists. In [21], it was shown that two facial analysis benchmarks are overwhelmingly composed of lighter-skinned subjects. A new facial analysis dataset balanced by gender and skin type was used to evaluate three commercial gender classification systems. It was concerning to see that darker-skinned females were the most misclassified group. Their error rate of up to 34.7% can be contrasted with that for lighter-skinned males, which was 0.8%. What this shows is that data may not represent all and therefore AI will not either. This problem is not unique to AI, and it has been noted that in different spheres of life, whether it is healthcare, medicine, safety, education, or employment, the data may not reflect diversity. The biases, prejudices, and follies of humans are in the data and therefore in the AI model; see [18] for an example in NLU. As an undesirable consequence, customers and end-users of AI have the risk of suffering discrimination based on their background, gender, income, etc. Research is going on to study if there can be solutions at the data collection stage which can mitigate the problem; see [28].

AI can make war much more lethal. Human history has been a record of weapons becoming more potent with time. With AI, the same pattern may continue. Centuries from now, students of history are likely to study how AI ushered in a new generation of lethal machines on the battlefield.

> *AI excels in carving manifolds and in fitting functions, but these manifolds and functions can be biased and flawed, having inherited the prejudices and imperfections of society.*

16.4   AI-Human Conflict

A classic Hollywood script is that of the evolution of AI into Super-AI which gets into conflict with its own creators, the outcome of which is not in favor of the latter. In imaginative stories, it is suggested that this occurs because AI develops the will to survive, flourish, and dominate or because it chooses to eliminate humans in order to create a better world. First consider the following script for a Hollywood movie.

"In the future, the Earth is invaded by extra-terrestrial AI machines of great power. Humans have no chance to defend against such a formidable opponent. A group of historians and scientists embark on a project with the goal of understanding the origin of the E.T. AI machines. When the truth is discovered, they are astounded. The machines are descendants of AI which was created by humans on the Earth in the twenty-second century to explore extra-solar planets. Due to their mission, they were programmed to be fully autonomous, that is, capable of functioning with zero human input. Nothing could interrupt and shut them down and they could create improved versions of themselves over time. After millions of years, they took on the task of creating a better universe. They classified humans to be an obstacle in the accomplishment of their goal and got into conflict with their own long-forgotten creators!"

If you find the above scenario imaginative, AI-human conflict can happen for a prosaic reason. It is a fact that it is exceedingly difficult to debug software completely such that we can handle every edge case and unseen scenario. AI can get into conflict with us because of flawed design in which its software gets into a rare state not foreseen by us. An unforeseen bug can make AI behave in an unpredictable, undesirable manner. These bugs can be even algorithmic because it is nontrivial to design safe objective goals and reward functions, which AI optimizes, with no undesirable and harmful side-effects, for all possible real-world settings. And, fully autonomous AI may redesign its own reward functions over time as in the above example.

> *AI machines will have their bugs and failure points. This would be a serious problem for AI designed to be fully autonomous.*

16.5 Choices Ahead

In 2015, theoretical physicist Stephen Hawking; entrepreneur Elon Musk; AI pioneering professors Geoffrey Hinton, Yann LeCun, and Yoshua Bengio; and several artificial intelligence experts signed an open letter on the topic of AI calling for expanded research on the societal impact of AI, giving many examples of research directions that can help maximize the benefits of AI. See [127].

There is a growing awareness in the AI community to make sure that the outcome of their work is ethical, fair, transparent, and accountable (see [13, 105, 144]), in response to the growing realization that AI technology and the AI field are not for everyone's benefits at the present time [239, 240]. Toolkits are being developed to detect and mitigate bias in AI models; see [12].

There has been debate among people about the future of AI. Some are optimists and see lots of benefits in AI. Mark Zuckerberg, Facebook's chief executive, is one of those who are not immediately getting alarmed at this time about potential pitfalls of AI. Others such as Stephen Hawking and Elon Musk have cautioned against the dangers. See [141] for this publicly debated topic.

The problem is not rooted in AI but in human nature. The founding fathers of the USA understood this very clearly (see Fig. 16.1) and knew that strong political institutions are the answer. A robust, vibrant, democratic political system which protects everyone based on checks and counter-checks and which allows freedom for beneficial progress will be needed. Choices we make today for the creation of such institutions for tomorrow's world will determine the future.

We need to ask whether the quest for super-smart AI is as important as the need to develop smart, shared, interconnected infrastructure for the common good. For example, why not invest in making inter-vehicle communication, traffic lights, roads, and street lights AI-enabled and interconnected, rather than in designing a

Fig. 16.1 Benevolent, ethical, transparent, and fair use of AI for the common good will be decided by institutions which are created and maintained by the people. Society will need to make choices for the road ahead. What goals do we consider important in the long run? How do we ensure that there is no adverse impact of AI? Would the shared infrastructure for public good benefit from AI? Would AI help other life-forms too? (Courtesy the US Capitol and Congress, Washington DC)

completely autonomous car with no infrastructure upgrades? The same goes for artificial general intelligence. Why not augment human intelligence and productivity with AI, rather than replacing humans and their jobs with machines? Autonomous cars and artificial general intelligence will be then just byproducts of research, but not what as a society we consider to be important collective goals. Economic shifts are inevitable because of automation. Worker retraining for new jobs has to accompany the adoption of new technologies and should be prioritized more highly.

We talked about AI which bring benefits to everyone. We owe our existence to the Earth, which supports an amazing variety of life. AI can be used to monitor the health of rainforests, which support life and refresh oxygen. It can be employed to ensure that marine life is flourishing. It can help different life-forms around the world as much it can help us. We must expand the definition of "we" and include other living things as beneficiaries of our science and technologies.

16.6   Chapter Summary

AI is a powerful technology which can be employed for great benefits and has immense potential to enhance the well-being of people. At the same time, like any other technology, one has to safeguard against the possibility that it can be used in ways which decrease well-being. In the future, there is a non-zero chance that advanced autonomous AI can get in conflict with humans due to unforeseen circumstances. AI depends on data, and therefore AI models inherit biases and flaws present in the data. There is a growing awareness in the AI community to develop practices which promote ethical, transparent, fair, and accountable AI. The long-term solution lies in strong political institutions which will ensure that AI is for everyone. Society has to clarify its expectations from AI and ensure that there is no adverse impact on people in the long term. Finally, all living things should be included in the definition of "everyone."

** Exercise 34   Ethical AI

1. How will you remove biases present in an AI model inherited from its training data? Give an example based on a literature survey.
2. In your opinion, which section of this chapter is a more realistic portrayal of the future: Sects. 16.1 or 16.3?

Chapter 17
Am I Looking at Myself?

I think consciousness will remain a mystery. Yes, that's what I tend to believe. I tend to think that the workings of the conscious brain will be elucidated to a large extent. Biologists and perhaps physicists will understand much better how the brain works. But why something that we call consciousness goes with those workings, I think that will remain mysterious. I have a much easier time imagining how we understand the Big Bang than I have imagining how we can understand consciousness ...

Edward Witten

Although you appear in earthly form, your essence is pure Consciousness.

Rumi

One of the direct, first-hand, and subjective experiences of our own intelligence we all have is when we say "Aha! I understand!" This phenomenal mental experience is referred as "qualia" by philosophers. Enjoyment of ice cream, of Mozart's music, of an orange sunset sky, of a fragrance-laden garden, and of a cool evening breeze has qualia. See [164].

What exactly do you mean when you say "I understand the proof" after your math teacher walks you through it? Suppose in the future, when we have developed a theorem proving AI model, will it be endowed with this subjective experience too?

Several thought experiments have been suggested by philosophers to understand consciousness and qualia; see [26, 164].

S. Dube, *An Intuitive Exploration of Artificial Intelligence*,
https://doi.org/10.1007/978-3-030-68624-6_17

1. Mary, the color scientist, who has been locked in an indoor colorless world, experiences colors for the first time in her life when she steps outdoors into a garden full of colorful flowers. The argument shows that she is able to solve the easy problem of objective color science while being disconnected from the hard problem of subjective visual perception of colors.
2. Can there be zombies who are identical to conscious humans at the molecular level but devoid of all phenomenal perception, qualia, and consciousness? This argument asks whether qualia is something in addition to this molecular arrangement in the physical universe.
3. In the Chinese room argument, one mechanically manipulates symbols without understanding the meaning of words. The conclusion of this argument is that programming an AI model may make it appear to understand language but it doesn't lead to any real understanding.

In this chapter, we discuss the topic of consciousness, which belongs more to philosophy rather than to science at this time, due to its connection to the topic of AI, which is inspired by the human mind, the seat of consciousness. The underlying question is whether there is a connection between intelligence and consciousness.

17.1   Is It Computable or Non-computable?

AI stands for *artificial* intelligence. Therefore, an AI model is an algorithm, that is, a Turing machine. The Church-Turing thesis asserts that the computation resources for AI consist of (1) an infinite memory tape and (2) a controller with finitely many states and finitely many symbols and a capacity to read and write symbols on the tape. No computation model can exceed the power of Turing machines. It is a computationally universal model.

In AI, we routinely formulate functions as

$$g : (X, \Theta) \to Y,$$

using real numbers. We don't care in our notation if the input X, the output Y, and the parameters Θ are uncountably infinite sets. We then immediately run into a fundamental problem of how to implement AI models in the form of computation. What does it mean to compute in the domain of real numbers which have infinite description? The Church-Turing thesis tells us to stay with finite descriptions such as rational numbers. Rational numbers are dense in the reals, so we do improve approximation as we increase computation time. What does it mean to have an infinite-resolution image of a cat? What does it mean when we say that the grayness value of a cat's pixel is a real number? That is disallowed by the theory of computation.

Of course, we are allowed to write our formulations in terms of reals and uncountable sets for convenience, but we should keep in mind that eventually we will have to work with a computer. Working with reals may facilitate conceptualization by using mathematical tools which allow us to think in terms of infinitesimally small changes and limits. Such AI models defined over reals will need to be approximated by computable models up to the desired accuracy. Any implementation of an AI model in this universe will have to honor the Church-Turing thesis.

If human intelligence is the result of an unknown, non-computable aspect of the universe, hidden in the microscopic fabric of space and time, transcendent to the Church-Turing thesis, then AI won't be intelligent in the true sense of the word, unless we can unravel the source of qualia and of the Eureka moment "Aha! I understand!" and are able to replicate it in a new revolutionary robot.

The above discussion is motivated by thoughts expressed by Roger Penrose in his thought-provoking book [160]. Also see [73, 162] to investigate this line of argument.

Roger Penrose suspects that consciousness supersedes AI. The former is non-computable and the latter is computable. Consider Kurt Gödel's incompleteness theorem, related to Alan Turing's undecidability results in the theory of computation, which asserts that if a mathematical system with its axioms and rules of inference has to be consistent, then it has to be also incomplete. There will be mathematical statements which cannot be proved or disproved. Gödel meticulously constructed the following statement in such a formal system:

"This statement is not provable"

which is self-referential. The statement is not provable or disprovable, because if it were, then we have a contradiction in either case. Is it true? Let's first ask an AI machine. An AI system can't infer anything because it works with axioms, computation steps, and logical rules, exactly like a formal mathematical system. To an AI system, only those statements which can be proved are true and those which can be disproved are false. The incompleteness theorem shows that there are true statements which can't be proved and there are false statements which can't be disproved. Therefore, an AI system can't reach the truth or falsehood of every statement.

Let us ask the same question ourselves. Is it true? You will just read the contents of the statement and know that it is indeed true because it is not provable. How did you establish the truth? You understood what the statement said. For you, it is more than a sequence of symbols to be manipulated by a logical inference system or a Turing machine. It has a meaning. One has to step out of the symbolic, mechanical, logic-driven proof machinery to see its truth by just semantically understanding what it is saying, which we are capable of doing. Your thought processes are likely to be – "Aha! That's what Gödel's incompleteness theorem says and that's what the statement says and I see its truth." You are like intuitive Captain James Kirk who outperforms logical Spock.

What is this subjective sense of understanding? Where is it coming from? Penrose believes that quantum mechanics, which is currently unfinished because it doesn't unify with the theory of gravity and because it has inexplicable measurement problem, can be extended in such a way that non-computational aspects will appear in the very microscopic fabric of the universe, where consciousness resides. There we "see" the truths, which AI will never be able to. This correlates with many mathematicians stating that their thinking is visual. This connection between the human mind, mathematics, and the universe is truly fundamental.

17.2   Is Consciousness Everywhere?

We belong to the physical universe, our mental states closely tied with physical states in our brains and our bodies, which constantly interact with the external environment. We are subject to the same laws as any other entity in the universe. Electrical charges and chemical processes provide the basis for brain processes which lead to thoughts, feelings, and awareness. How can qualia emerge from something which is not conscious? We are 60% water. The human brain is 73% water. Electrical charges run through neurons in our glucose-fed, wet brain. It is the same dance of molecules, atoms, electrons, and quarks in our brains as in the ocean waves, in rain and lightning, in solar flares, and in the birth of stars. In microcosm and macrocosm, we share the same physics as any other object in the universe.

Let's say there is a consciousness function C which classifies a region of the universe as conscious or not. Let's say that

$$C(m) > 0,$$

$$C(s) = 0,$$

for the human brain m and for pretty much any other part s of the universe. For dolphins, elephants, and a few other animals whom you deem to be conscious, feel free to make C non-zero. Can such a strange function really exist, when m is identical to so-called unconscious things at the molecular level? Will it be more realistic to say that

$$C(s) \geq \epsilon$$

where ϵ reflects a primitive form of the qualia? What will be the consciousness of a robot?

It is intuitive to connect the seat of consciousness with the brain. In fact, two current theories of consciousness are global neuronal workspace theory and integrated information theory (see [114]), and both theories abstract out certain features of the brain. The former postulates that consciousness is a result of the particular way a global, brain-scale, and coherent processing occurs in the brain. The latter postulates that the underlying physics of the brain mechanism is the root cause which endows it with high intrinsic causal powers. The more the current state of a system specifies its past and its future, the more causal power it has.

Modern physics has shown that not everything is what it seems. Matter is not solid stuff but evolving waves of possibilities which interact with each other and create reality. In the space-time fabric, it is all a mysterious dance of relationships, interactions, and transformations. If qualia emerges out of the same processes, it must have roots in more fundamental properties. It can't emerge out of nothing. This is panpsychism [163], the belief that the source of consciousness is everywhere and goes all the way to the fundamental ingredients of the universe, because everything is connected to everything. Therefore, we can't rule out the possibility that in the distant future robots will achieve human-like consciousness by tapping into this source of qualia. After all, we are ourselves biochemical robots designed by nature. Why would a biochemical robot be conscious while an electronic or some other kind of robot would not be?

17.3 Who Is the Storyteller?

We can state mathematically the evolution over time of the state of the universe s_t,

$$s_t = F(s_{t-1}),$$

and of our own brain m_t, which is embedded in it,

$$m_t = L(m_{t-1}, e_t, b_t),$$

where e_t is the state of the environment and b_t is the state of our bodies. The functions F and L are governed by scientific laws and F includes L (Fig. 17.1).

Function F subsumes Darwin's theory of evolution. Mutations, reproduction, genotype-to-phenotype translation, the changing environment, and survival of the fittest are all physical processes, and therefore all are part of F. These processes don't care about the accompanying qualia of our mental states. Qualia is a subjective experience, while the laws governing the universe are objective and *qualia-agnostic*. Time passes and evolution occurs like an iterative algorithm. Function F does not need to be qualia-aware.

Mental states m_t are part of the global physical state s_t. Evolution will occur even if the qualia was absent. But we know that qualia is there. It is our direct experience of everything unless we are fooled somehow. It is as if there is another function,

$$q_t = Q(m_t),$$

which creates qualia. That is, the physical processes get accompanied with their conscious experience. Alternatively or perhaps equivalently, consciousness is nothing but exactly the same underlying reality, which outwardly appears as the qualia-agnostic observable physics, while being experienced inwardly by us in a qualia-rich unobservable way.

Fig. 17.1 Our rich subjective experience of life and our strong self-awareness constitute the greatest mystery of the universe. This can't be explained by Darwin's theory of evolution, which uses objective scientific laws which govern the evolution of the universe in space-time, because it doesn't need qualia. The physics of the universe is qualia-agnostic. It is not clear whether we will ever be able to endow AI with qualia. This is the hard problem of consciousness. At the same time, AI is likely to pass and exceed the Turing test, and we may solve the easy problem of consciousness. The images show that all of us go through life which is experienced as a meaningful story. The left picture is of the author's late mother, c. 2000, symbolizing her self-awareness as well as his subjective experience of life. What is the source of this universal experience? Did the robotic boy David in the movie A.I. (2001) feel the same way? (Image credit: artist Abraham Bloemaert (1566-1651), courtesy the National Gallery of Art, Washington DC)

Our lives are stories of seasons following seasons and of joys and sorrows which accompany these. Who is the storyteller who consistently creates the qualia of joy at the birth of our children and of grief when we lose loved ones? As years roll by, we grow older living these stories.

We should solve the easy problem of AI by creating machines which pass and outperform the Turing test of intelligence. There is already enough work in the easy problem to keep us busy for centuries.

> *We may solve the easy problem of consciousness, which entails understanding the science behind brain processes and which will inspire future work in AI. The hard problem of understanding the science behind qualia and that of endowing machines with it may remain out of our grasp forever.*

17.4   Chapter Summary

In Bicentennial Man (1999) and A.I. Artificial Intelligence (2001), Hollywood endowed advanced futuristic robots with consciousness. In this chapter, consciousness was examined by employing the concept of qualia from philosophy and tools from mathematics and the theory of computation. Arguments by Roger Penrose make use of Gödel's incompleteness theorem to show that there is a dimension to human intelligence which can't be implemented in AI. The implication for AI is that AI will have to honor the Church-Turing thesis since its power will be limited by that of a Turing machine and of a formal logical system. Current theories of consciousness abstract out certain features of the human brain in order to develop an objective criterion to quantify consciousness. Panpsychism, one of the philosophies of mind, views consciousness to be a fundamental ubiquitous property of the universe. Though we have made progress in understanding the universe and the evolution of life, it is not clear why there is consciousness which accompanies these processes. It is likely that the problem of consciousness will remain intractable and outside the scope of AI.

> ## * Exercise 35   Types of AI
>
> Define narrow, weak, and strong AI giving an example for each; see [162].

Appendix A
Solutions

> It's what you learn after you know it
> all that counts.
>
> John Wooden

Answer of Exercise 1

The result follows from the fact that any open subset of the Euclidean space is a manifold. In general, any open subset of an n-dimensional manifold is itself an n-dimensional manifold. Under the assumption that for any image x of a given class, there is room for slight perturbation ϵ which preserves the class, we can formalize this as an open neighborhood $B(x, \epsilon)$ centered at that point. Therefore, the given image class is an open subset embedded in the surrounding Euclidean space. See Fig. A.1 for an open subset of $\mathbb{R}^2$.

Answer of Exercise 2

For a literature survey, we refer to [253], which applies tropical geometry, a branch of algebraic geometry, to count the number of linear regions. In tropical geometry, the multiplication operation in classical polynomials is replaced by addition, and the addition is replaced by maximum. Such a polynomial will be a convex, piecewise linear function. The ratio of two polynomials, referred to as tropical rational map

© The Author(s), under exclusive license to Springer Nature Switzerland AG 2021
S. Dube, *An Intuitive Exploration of Artificial Intelligence*,
https://doi.org/10.1007/978-3-030-68624-6

Fig. A.1 An open subset of
the Euclidean space

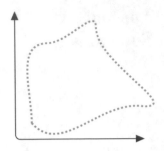

or tropical quotient, will be a non-convex, piecewise linear function. It will be the standard difference of two tropical polynomials, since multiplication corresponds to sum.

A remarkable result follows. The family of functions defined by deep feed-forward neural networks with ReLU with integer weights is exactly the family of tropical rational maps.

Note that restriction to integer weights does not really lead to any practical loss of generality as real weights can be approximated by rational numbers, which can be scaled up to turn them all into integers, resulting in scaled output.

With the tropical geometrical interpretation, one can count the number of linear regions and show that it increases polynomially with the width and exponentially with the depth. See also prior work in [145, 171].

Answer of Exercise 3

For regression, mean squared error (MSE) is a popular loss function. It is the mean of the squares of the differences of the ground-truth values and the predicted values, that is,

$$\frac{1}{N}\sum_{i=1}^{N}(y_i - \hat{y}_i)^2,$$

and minimizing MSE is equivalent to minimizing the L^2 loss because

$$\text{MSE} = (L^2 \text{ loss})^2 / N.$$

It is sensitive to outliers, and therefore for certain applications, we use the L^1 loss. We mention mean absolute deviation (MAD),

$$\text{MAD} = (L^1 \text{ loss}) / N.$$

N does not play a role in minimization, and we could have minimized the sum of squared errors (SSE) and the sum of absolute deviations (SAD).

The L^p loss, where $p \geq 1$, for linear regression is a convex function. For the L^1-norm case, it is non-differentiable convex optimization. Another alternative is Huber loss. Huber loss is in between the L^2 and L^1 losses. It is similar to the L^2 loss when the error is small; otherwise, it is like the L^1 loss, making it more robust to outliers than the L^2 loss.

For binary classification, for a positive sample, the loss is $-\log p$, where p is the predicted probability, and for a negative sample, it is $-\log(1-p)$. Let the ground truth for a sample be $y_i \in \{0, 1\}$, where y_i is 1 for the positive labels; else it is 0. Then, the cross-entropy loss for a sample is

$$\mathcal{L}(x_i) = -y_i \log p_i - (1 - y_i) \log(1 - p_i).$$

This is a convex function for logistic regression. Interestingly, the L^p loss will be non-convex for logistic regression. There is an interpretation of the cross-entropy loss in terms of the Kullback-Leibler divergence, also known as relative entropy, which measures how far apart two probability distributions are with respect to one another. Given two probability distributions P and Q,

$$\text{Cross-Entropy}(Q \| P) = \text{Entropy}(Q) + \text{KL}(Q \| P).$$

For $P = [p, 1 - p]$ and $Q = [q, 1 - q]$, the KL divergence is

$$-q \log p - (1 - q) \log(1 - p) - (-q \log q - (1 - q) \log(1 - q)).$$

For logistic regression, Q is the discrete ground-truth probability distribution $[q, 1 - q]$, its entropy being a fixed value. Q is $[1, 0]$ for the positive samples and $[0, 1]$ for the negative samples. The entropy of Q is zero for the special case when $q \in \{0, 1\}$,

$$-1 \log 1 - 0 \log 0 = 0.$$

Therefore, minimizing the cross-entropy,

$$-q \log p - (1 - q) \log(1 - p),$$

is the same as minimizing the KL divergence. It is easy to generalize the loss for K multiple classes, where P and Q are K-dimensional. The ground-truth probability of a sample is a one-hot vector Q, with 1 being for the correct class and 0 being elsewhere. We add up these losses over a set of samples. The total loss over the training set x is

$$\mathcal{L}(x) = \sum_{x_i \in x} \mathcal{L}(x_i).$$

The sum of convex functions is convex. Therefore, the total loss is a convex function.

Answer of Exercise 4

If there exists some room around each feature vector for slight perturbation which does not change the image class, then feature regions will satisfy the manifold property.

Answer of Exercise 5

Speech and Audio   A sound wave is a function of time. Let N be the temporal resolution, that is, we are given values of the function at N uniform ticks of time. There will be slight room to perturb any waveform without changing its class. Therefore, for speech and audio waveforms, we should have manifolds in $\mathbb{R}^N$ space.

NLU   In NLU, we have a vocabulary, and there are rules of generation directed by syntax and semantics. Objects of interest such as articles, stories, and poems are not objects in $\mathbb{R}^d$ for some d and are qualitatively different from images and speech. Words are discrete entities and there are finitely many of them. The generative parameters are hard to formulate. They exist in realm of thoughts, emotions, and the world model. Do thoughts and emotions vary continuously or are they inherently discrete? What is the connection between words and the speakers' world view or perceptions? Should the words have discrete token-based representations or numerical vector representations? We consider the latter approach.

We can work in $\mathbb{R}^d$ space and express natural language entities such as words in this space as points. This is the approach of word embeddings, a special case of distributed representation. A distributed representation of a word is the activation pattern of neurons of an AI model. Once we are in the Euclidean space, we can apply similar techniques as for images and speech to solve practical problems. The distributed representation of words leads to interesting questions. What do points belonging to the gap between two valid words such as happiness and sadness mean? Are these semantically valid concepts even though we may not have any dedicated words for them in the vocabulary? Text documents such as stories are more than words. They require a world model, subjective experience of emotions, abstract concepts, and flow of time. How to capture these intricate multi-layered and multi-faceted dynamical structures in rich numerical representations? The goal of AI is to be able to create distributed representations of such structures which are close to their semantics.

It seems we cannot model natural language in terms of manifolds in an obvious, straightforward manner. Yet, we do that routinely in practice for certain problems. Say we embed words in $\mathbb{R}^d$. Consider documents consisting of exactly N words. Then, each document is a point in high-dimensional space $\mathbb{R}^{dN}$. In applications, we often build a compressed, fixed-size representation of variable-size documents. We can approximate groups of similar documents as manifolds and use neural networks to carve them out. Here embeddings are like colors of words analogous to colors of image pixels.

Software Programs   This is very similar to the NLU case above. For malware programs consisting of N bytes, slight perturbation of any program may result in an invalid program which does not execute on the computer. Similar to text documents, there may be good d-dimensional embeddings of bytes which capture their semantic similarity. Malware programs in a suitably learned distributed representation space can be approximated as a manifold by a neural network. Hopefully, this manifold will have both malware programs and invalid programs but no benign program.

Answer of Exercise 6

1. The derivative is $p - G$. This can be seen as the product of the following:

$$\frac{dp}{dy} = p(1 - p),$$

$$\frac{dL}{dp} = \frac{-G}{p} + \frac{1 - G}{1 - p} = \frac{p - G}{p(1 - p)}.$$

2. It may just happen that for the particular training dataset, the parameters always get values during the backpropagation algorithm which make the neuron's carving cut totally outside the training data. For all training examples, SGD may fail to wake up the neuron if the carving cut remains always outside.

3. No, loss still depends on what the AI model predicts for the negative classes. The probabilities p_is are obtained by the softmax function, which is dependent on outputs of the model for all classes. Therefore, the backpropagation algorithm propagates the gradients along the entire computation graph. In fact, the derivative of the loss with respect to the i-th input value y_i to the softmax layer,

$$p_i = \frac{\exp(y_i)}{\sum_j \exp(y_j)},$$

is $p_i - G_i$, which is exactly the same expression as for the special sigmoid case above.

Answer of Exercise 7

In CNN, we should have a similar design of artificial neural networks which have analogs of simple and complex cells, that is, low-level simple neurons which respond to a particular feature in a particular region of the image and higher-level complex neurons which have translation spatial invariance.

It was a serendipitous discovery by Hubel and Wiesel in one of their initial experiments to make cat neurons fire in response to an oriented line or edge because initially they were not having luck with other image features such as circles. It was an accidental response of the cat neurons to the edge of a piece of glass in the projector which they were moving which made them succeed.

Why such a strong response to the edge of the piece of glass? Due to the hard work of neuroscientists, we know that a simple cell selectively responds to an image feature which has a specific spatial localization and a specific spatial orientation and which occurs at a particular spatial scale (bandpass).

Answer of Exercise 8

The paper shows that an unsupervised learning algorithm does lead to neurons which respond to localized, oriented, and bandpass image features. The goal was to build a dictionary of basis functions by applying an iterative optimization technique which combined the reconstruction goal with the sparsity constraint. The basis functions which the authors got using this totally unsupervised learning algorithm remarkably had similar image feature properties to what neuroscientists already knew.

It is like software simulation of the evolutionary process which has given rise to these simple cells in the visual cortex over hundreds of millions of years. The gradient descent steps in the iterative optimization algorithms are evolutionary steps over generations of organisms trying to move the visual cortex towards the goals of reconstruction and sparsity. To see this, let us put ourselves in the shoes of biological evolution. What kind of mammalian primary visual cortex would evolution favor in the design of simple cells? Consider two choices for the number of neurons which fire when eyes looks at an image patch of interest.

1. Lots of simple cells respond.
2. Very few simple cells respond.

The first choice would be very taxing on our brains. It will be good to have very few neurons respond in order to conserve energy. This is the sparsity criterion. At the same time, in order to see well and to avoid missing any image details in the patch of the scene our eyes are focusing on, the image features which these few neurons respond to should capture all the image information. From the selective image features of the neurons, we should be able to recreate the image patch. This is the reconstruction criterion. The authors combined the two criteria.

Answer of Exercise 9

1. Figure 1.3 shows a carving by a feed-forward neural network with one hidden layer. There are two input neurons, three hidden neurons, and one output neuron. The three straight dashed lines are made by the hidden neurons. The output neuron makes the dashed bending line.
2. An alternative way to look at the working of an AI model is by noticing that it distorts the input in the feature spaces. Therefore, a straight line segment in the input space will get bent in the feature spaces. The higher the level of a feature space, the more the bending. An alternative method could be to measure the distortion of the input space, by measuring how much a line segment bends and twists around in the highest feature space computed by the last layer.

Answer of Exercise 10

The big-O, Ω, and Θ notations are used in computer science to characterize the growth of the running time of algorithms with respect to the input size, and they can be applied to characterize the growth of any function in general. If we have

$$f(n) = O(g(n)),$$

then it means $g(n)$ is an upper bound of $f(n)$, that is, there exist some positive constants n' and c such that $f(n) \leq cg(n)$ for all $n \geq n'$. Ω-notation is for lower bounds, and Θ-notation is for both upper bounds and lower bounds.

For a binomial coefficient $C(n, k)$, for fixed k,

$$\binom{n}{k} = \frac{n(n-1)\ldots(n-k+1)}{k!},$$

is a polynomial of degree k, and therefore it is $\Theta(n^k)$. Note that term containing the highest power has a positive coefficient. The number of summands in $r(n, d)$ is $d + 1$, a fixed number which does not depend on n. The whole sum is a polynomial of degree d, and it is bounded by the term containing the highest power d which has a positive coefficient.

Answer of Exercise 11

See Sect. 3.1.2 to compute k. Suppose the image is $H \times W$ and the convolutional filter is $n \times n$. Let B be the padding width at the image border to take care of convolution at the border. Let the convolution stride be S. Then, the number of points in the sliding window space is

$$k = ((W - n + 2B)/S + 1) \times ((H - n + 2B)/S + 1).$$

Answer of Exercise 12

The loss landscape changes as

$$\frac{1}{2}(\lambda_1 \cos^2 \phi + \lambda_2 \sin^2 \phi).$$

To see this, express the unit-length vector u in the direction making an angle ϕ with v_1 as

$$u = v_1 \cos \phi + v_2 \sin \phi,$$

where we use v_1, v_2 as the orthonormal axes. The result follows by expanding

$$u^T H u = (v_1^T \cos \phi + v_2^T \sin \phi) H (v_1 \cos \phi + v_2 \sin \phi),$$

and using the fact that v_1 and v_2 are perpendicular eigenvectors. That is, their dot product is zero,

$$v_1^T v_2 = v_2^T v_1 = 0.$$

We also use the fact that, for $i \in \{1, 2\}$,

$$H v_i = \lambda_i v_i,$$

and that v_i is of unit length, so the self dot product is 1,

$$v_i^T v_i = 1.$$

The sign and magnitude of an eigenvalue determine whether moving along the corresponding eigenvector will make the loss increase or decrease and by how much. When we walk somewhere in the middle of v_1 and v_2 on the optimization landscape, the change is in the middle too.

Answer of Exercise 13

In [122], it is argued that noisy updates can help the AI model by keeping it from getting stuck in a local minimum. Though the argument is given for local minima, it should generalize to any critical point. Batch gradient descent does have the advantage of full convergence, and it is easier to do its theoretical analysis. SGD may stall and fluctuate around a good local minimum. Batch gradient descent is slower because of redundancies of patterns present in the training data.

The second-order methods based on the Hessian matrix are not used in AI because the Hessian is an $N \times N$ matrix and these methods become impractical for large optimization problems. The time complexities of the Newton, the Quasi-Newton (BFGS), and the Levenberg-Marquardt methods are $O(N^3)$, $O(N^2)$, and $O(N^3)$, respectively.

Answer of Exercise 14

The learning rate has a big influence on the speed of training. By reducing the learning rate, you can often speed up the training as you get closer to a local minimum.

Answer of Exercise 15

Let D be a training data of n samples $x_1, x_2, \ldots, x_n$ with the ground truth,

$$y_i = f(x_i) + \text{Irreducible Error},$$

where $f(x)$ is the true, golden function. Let $\hat{f}(x)$ be the predicted value. Let x be a test sample. Then, take the expectations over training datasets D of size n,

$$\mu = \mathbb{E}_D[\hat{f}(x)],$$

$$\text{Bias} = \mu - f(x),$$

$$\text{Variance} = \mathbb{E}_D[(\hat{f}(x) - \mu)^2],$$

$$\text{Total Error} = \text{Bias}^2 + \text{Variance} + \text{Irreducible Error}.$$

Answer of Exercise 16

Cosmological parameters in theories such as LambdaCDM are generative parameters, because along with other ingredients of space, time, and energy, one can create entire universes, and we are fortunate to be living in one where these parameters have the right values! In order to estimate these constants, we can do the following.

1. **Generative Method:** We choose the values which make the actual observed data most likely. This is the maximum likelihood estimation in artificial intelligence. We can follow two techniques.
   Slow Brute Force Method: To compute the likelihood, we will generate many universes through computer simulations and score how well they match with our actual universe to compute their likelihood. We have two loops, one nested in the other. In the outer loop, we vary the cosmological constants, and in the inner loop, by varying the initial conditions of the universe, we simulate universes. We can then compute for different cosmological constants how likely we are going to see the things observed in our universe over different initial conditions.
   Faster Simplex Method: In the brute force method, the number of simulated universes will be huge; therefore, we can compute the likelihood based on a small set of simulations as starting points and use an uphill simplex method to tweak the parameters for more simulations to maximize the likelihood. In the simplex method, an n-dimensional tetrahedron flip-flops upwards on the optimization landscape. We can't use the gradient descent algorithm because the likelihood function is based on simulations and is not differentiable.
2. **Predictive Method:** Making use of a generative model to create universes in simulations is a nice idea, but can we come up with a faster method which computes similarity-based likelihood? We can build a supervised machine learning model, for example, a deep neural network, which goes in the reverse direction. It takes as input a universe with its distribution of galaxies and dark matter and outputs the values of the cosmological parameters. We train this neural network on the simulated universes, and finally we apply it on the actual universe we live in.

See [157, 158, 173] for applications of ML and AI to cosmology.

Answer of Exercise 17

The proof is based on the how we run the Markov chain Monte Carlo algorithm on Θ and from the correctness of the MCMC algorithm. The MCMC algorithm does not require an exact probability distribution to be known, but only the ratio of probabilities of any two samples. In the limit, the algorithm draws samples as given by P.

Answer of Exercise 18

The standard GAN is based on the Jensen-Shannon divergence, which computes the discrepancy between two probability distributions. It is a symmetric divergence derived from the Kullback-Leibler divergence. Precisely speaking, if the discriminator D has been trained optimally, then the generator G minimizes the JS divergence. In [4], it is shown that the Wasserstein metric, which is also known as the Earth Mover Distance, is continuous and the Jensen-Shannon divergence is not. Moreover, for two non-overlapping distributions, the JS divergence fails because moving the two closer doesn't decrease the JS divergence. Clearly, the EMD metric works fine. That is why one can learn probability distributions better with the Wasserstein metric using gradient descent.

Answer of Exercise 19

Compared with Q-learning, Sarsa tends to be conservative as it seeks to avoid negative-reward actions while exploring, and it finds a near-optimal policy. If you are trying to navigate a world full of unknown dangers, then Q-learning will choose an optimal policy which brings you closer to these dangers and still maximizes the expected reward obtained by reaching treasures, whereas Sarsa will find a near-optimal path keeping you away from the dangers. This happens because for Sarsa, the behavioral policy is the same as the learned policy, whereas for Q-learning, the two are different. Therefore, Q-learning is algorithm B.

Answer of Exercise 20

An improvement is made by using double Q-learning to fix the problem of overestimation by DQN and to improve its performance; see [77]. In the maximum operation,

$$\max_{a'} Q(s', a'),$$

we use a DQN to find the best action,

$$a_m = \max_{a'} Q_1(s', a'),$$

and use a second DQN to predict the value in the computation of the TD error,

$$r + \gamma Q_2(s', a_m) - Q_1(s, a).$$

Answer of Exercise 21

We are given a constant function with added noise,

$$x = \theta + \epsilon,$$

where ϵ is the noise. Then the maximum likelihood estimate of θ given observations $D = \{x_1, x_2, \ldots, x_N\}$ is

1. $\hat{\theta} = \text{mean}(D)$ if ϵ is Gaussian.
2. $\hat{\theta} = \text{median}(D)$ if ϵ is Laplacian.

Alternatively, if we want to fit a constant function to D,

1. The mean will minimize the L_2 error. Equivalently, we minimize the mean squared error.
2. The median will minimize the L_1 error. Equivalently, we minimize the mean absolute deviation.

Here is an outline of the proof. Consider the probability distribution of the given stochastic process,

$$k_1 \exp(-k_2 |x_i - \theta|^\alpha)$$

for some positive constants k_1 and k_2. For the Laplacian function, $\alpha = 1$. For the Gaussian function, $\alpha = 2$. The likelihood of the observed data is maximized when the product of the exponential terms is maximized. This happens when the negative logarithm of the product is minimized. The negative log-likelihood is minimized when

$$\sum_{i=1}^{N} |x_i - \theta|^\alpha$$

is minimized. It can be easily shown that the mean minimizes the sum of squared deviations and the median minimizes the sum of absolute deviations.

Answer of Exercise 22

1. All are convex, except for neural networks.
2. SVM has a single hyperparameter known as the slack parameter which determines the total amount of error which is allowed. It uses a loss function called as the Hinge loss.

3. The split children nodes should have better class separation than their parent node. One uses a mathematical measure of the purity of a class such as Jini or entropy.

Answer of Exercise 23

The numbers of parameters in MobileNet-V2, ResNet-50, ResNet-152, Inception V3, VGG-19, and NASNet-Large are 3.54, 24.64, 60.38, 23.85, 143.67, and 88.95 millions, respectively. The model size will be four times as many in MB, assuming 4 bytes are needed for each floating-point parameter.

Answer of Exercise 24

The ROC curve is a plot of TPR versus FPR. For FPR, we need to compute the total number of labeled negatives. Suppose we are given a ground-truth bounding box B for a cat in an image. Consider an IOU threshold of T. Any bounding box in the image whose IOU with B is less than T is a negative example. Since the number of such negative examples is extremely large and it is inconvenient to even compute it, the PR curve is used. Of course, in theory, the ROC curve does exist for the problem.

Answer of Exercise 25

An effective approach will be to formulate the problem as a density map estimation problem. The output of the AI model is a density map which is summed across spatial locations to obtain an estimate of the size of the crowd. This is more robust than formulating the problem as an object detection problem, in which one detects and counts, or as a semantic segmentation problem. The ground truth is a sparse binary map. The prediction is a real-valued density map. One possibility is converting the ground truth into a density map with a Gaussian blur operation and then using the Kullback-Leibler divergence or Jensen-Shannon divergence.

In [220], the optimal transport theory is applied, which measures the distance between two distributions by computing an optimal plan to move probability mass from one to the other. It minimizes the cost of moving this infinitely fine sand from one pile to the other over all possible transportation plans. This minimal cost is called the Earth Mover Distance when one is working with histograms, and it is also called the Wasserstein distance which is used in GANs. In [220], a suitable optimal transport loss function is designed based on the Monge-Kantorovich distance.

When we have empirical distributions given by N observations, the Hungarian algorithm can be employed to obtain the optimal solution of a given mass transportation problem. A faster, approximate solution can be obtained by the Sinkhorn algorithm which discretizes the space [220].

Answer of Exercise 26

We refer to [218]. At the highest level, self-attention is given by

$$\text{Attention}(Q, K, V) = \text{softmax}\left(\frac{QK^T}{\sqrt{D}}\right) V.$$

Consider a sequence with N words,

$$\{w_1, w_2, \ldots, w_N\}.$$

Each word w_i has an embedding given by $E(w_i)$. For each word w_i, obtain the query, key, and value from its embedding $E(w_i)$ using three learned functions $Q(w_i)$, $K(w_i)$, and $V(w_i)$. Compute the attention score vector for w_i by taking dot products of its query with the key values,

$$A(w_i) = \text{softmax}(K \times [Q(w_i).K(w_1), Q(w_i).K(w_2), \ldots, Q(w_i).K(w_N)]),$$

where $K = 1/\sqrt{D}$ and D is the dimension of the query and key vectors. The normalization factor K makes the model numerically stable. Let

$$A(w_i) = [a_1, a_2, \ldots, a_N].$$

Compute a weighted combination of the value vectors,

$$E'(w_i) = \sum_j a_j V(w_j),$$

which is the encoded representation of w_i computed by the self-attention layer. The process is repeated for the next layer with the new embeddings E' as the input.

Answer of Exercise 27

A path in the probability matrix M going from left to right is a candidate transcription whose probability is the product of the probabilities of all the graphemes along the path. Taking the logarithm, it is the sum of the log probabilities. Note that a path

aaawwww⟨b⟩esso⟨b⟩mme

evaluates to "awesome" by squeezing out blanks and compressing repeating graphemes not separated by blanks. Let this operation be denoted by function h, for example,

$$\text{"awesome"} = h(\text{aaawwww}\langle b\rangle\text{esso}\langle b\rangle\text{mme}).$$

Then, the total probability of the transcription "awesome" is the sum of the probabilities of all paths p which evaluate to the word,

$$\text{Pr("awesome")} = \sum_{h(p)=\text{"awesome"}} \text{Pr}(p),$$

which can be computed by dynamic programming. The gradient of the probability can be computed and used to train the AI model using backpropagation.

Answer of Exercise 28

For the standard GAN, the discriminator loss is

$$-\mathbb{E}_{y}[\log D(y)] - \mathbb{E}_{\theta}[\log(1 - D(G(\theta)))],$$

where $\mathbb{E}_{y}$ is the expectation over the real images and $\mathbb{E}_{\theta}$ is the expectation over all possible generative parameters (random vectors). For the translator GAN, the loss is conditioned over an additional input x. For paired-image translation tasks, this additional input is the source image x. For multi-class generation, x is set to the class label. These are called conditional GANs.

For image translation, the discriminator loss is

$$-\mathbb{E}_{x,y}[\log D(x, y)] - \mathbb{E}_{x,\theta}[\log(1 - D(x, G(x, \theta)))],$$

where $\mathbb{E}_{x,y}$ is the expectation over the translated images, which are provided as the ground truth, conditioned over x, and $\mathbb{E}_{x,\theta}$ is the expectation over all random noise vectors conditioned over x. The discriminator seeks to distinguish fake translation image pairs from real ones, when given the pairs of source images and the corresponding translated images.

We want the generator to be an autoencoder; therefore, we add the L^2 reconstruction loss to the generator loss function,

$$\mathbb{E}_{x,\theta}[\log(1 - D(x, G(x, \theta)))] + \lambda L^2(y, G(x, \theta)).$$

Therefore, a translator GAN ensures not only that the translations by the autoencoder-based generator have low errors but also that these translations are indistinguishable from real ones. GAN-based translation is an enhancement of autoencoder-based image translation.

A couple of remarks are due at this point. It has been empirically observed that the noise vector θ is no longer effective in adding stochastic variations to the output and can be removed altogether. Noise can be more effectively added by having dropout layers in a few upsampling layers of the decoder part of the generator. Furthermore, the L^1 reconstruction loss could offer less blurring, especially in the case when multiple reconstructions are valid. This stems from the fact that the L^2 loss will aim for the mean reconstruction and the L^1 loss for the median reconstruction. For details, see [101]. Finally, one can remove the logarithm operation and convert it into a Wasserstein paired translator GAN.

For the cycle GAN, one has two generators F and G and two discriminators D_X and D_Y. Total cycle consistency loss is

$$L = d(X, F(G(X))) + d(Y, G(F(Y))),$$

where d is the L^1 distance. The cycle reconstruction loss is combined with the standard reconstruction loss. For the generator G, the loss is

$$d(Y, G(X)) + L,$$

and for the generator F,

$$d(X, F(Y)) + L.$$

The discriminator D_X learns to distinguish between X and $F(Y)$. The discriminator D_Y learns to distinguish between Y and $G(X)$. The loss is the binary cross-entropy loss. See [257] for details. Note that this is an unpaired translation task in which one is given two sets of images, for example, daytime images and nighttime images. Therefore, if one had just used the standard reconstruction loss, it would not have been meaningful. The cycle consistency loss imposes additional constraints to make the model learn correctly and to enforce geometrical consistency on the translation of the objects.

For both the paired translator GANs and unpaired cycle GANs, it is preferable to use a CNN with skip connections, such as U-Net [180], to allow information from different scales to flow from the encoder to the decoder within the generator.

Answer of Exercise 29

Though static computation graphs are faster to execute, dynamic computation graphs offer certain key advantages. One of the biggest advantages is that complex

AI models can be more easily implemented. There have been examples in NLU which illustrate this strength of dynamic computation graphs and which are very cumbersome with static computation graphs. In addition, the code is easier to debug and looks more like a regular Python code.

Answer of Exercise 30

The minibatch size is B. Suppose one needs 4 bytes to store one floating-point value. For inference, one needs to store the N activation values for the forward pass, and therefore $4 \times B \times N$ memory is needed, where typically B is small. For training, one needs to compute the gradients with respect to the M learnable parameters and with respect to the N activation values. Therefore, one needs

$$4 \times B \times (2N + 2M)$$

bytes if one is using SGD without momentum. Typically the activation memory of $2N$ is more than the optimizer memory of $2M$. Since momentum is frequently used, one needs to keep the past gradients, and the memory requirement increases to

$$4 \times B \times (2N + 3M)$$

bytes. For RMSProp [62], which needs to store the first and second moments, it will increase further to

$$4 \times B \times (2N + 4M).$$

Since the GPU memory is limited, it constrains the largest possible value of B.

Answer of Exercise 31

We refer to a Harvard Business Review article; see [53]. According to this article, the global economic impact of AI is estimated to be $13 trillion over the next decade. At the same time, only 8% of organizations are adopting core practices which promote widespread adoption of AI. Most of the time, ad hoc pilots are run, or there is limited single application of AI. The root cause is inability to evolve the organization into a form which is amenable to AI adoption.

Answer of Exercise 32

This is an interdisciplinary exercise, and the best outcome will be achieved by students with diverse backgrounds such as computer science, business, and mechanical engineering working together. Taking AI to the real-world requires much more than technical expertise and is an example of team effort. Here is a use case to get you started along similar lines. The use case has to be converted into a concrete business plan backed by actual numbers and graphs.

Business Objective Develop an automated technological solution to accurately and efficiently diagnose malaria in patients through AI-based image analysis of blood smears.

Markets

1. The endemic markets (e.g., South Asia, Africa)
2. The non-endemic markets (e.g., N. America, Europe)

Background Malaria is a leading cause of death in tropical and sub-tropical countries with an estimated 1–2 million deaths every year. Early diagnosis of malaria in patients is critical in order to save their lives. The gold standard to diagnose malaria is visual microscopic analysis of thin blood smears stained with Giemsa. Unfortunately, this visual analysis is an extremely tedious, time-consuming process which depends on human expertise. In the non-endemic markets, technicians lack expertise due to lack of practice. In the endemic markets, due to the sheer volume of the suspected malarial cases and the tedious nature of the job, visual analysis is not performed to its completion, and the patient often gets diagnosed based on clinical symptoms. We will build an automated computer-based solution which makes use of advanced AI-based computer vision techniques to accurately and efficiently diagnose malaria.

Risks and Favorable Factors In the past, people have attempted to build digital image processing systems for malarial diagnosis, and there are a few research publications in this area. Commercial systems have not been built in the past due to lack of focused advanced work giving robust results.

There are four kinds of malarial parasites, which can have different growth stages. Furthermore, there could be variations in images of blood smear slides prepared by different technicians and different clinics. In the endemic markets, there are risks associated with capital costs, lack of electricity, inadequate ruggedness of the instrument for field use, maintenance problems, and insufficient training of technicians to calibrate and use the system.

There are two favorable factors which increase the chances of success. First, there have been significant advances in computer vision using AI (deep learning). Second, hardware costs are falling rapidly.

Phase 1 In the first phase, we will target the non-endemic markets. A state-of-the-art digital camera coupled with a microscope will be used to acquire images. A motorized stage to move the slide will be attached to the digital microscope. The instrument will be connected to a laptop or a desktop computer through USB.

The computer will run AI software written in Python using Keras. Target cost of the system: $3000 to $5000. Time to develop the prototype is 6 months. Funding and team required:

1. Seed capital of $0.5 M.
2. Consulting services of a digital camera/microscope image acquisition expert and a mechanical engineer to build a prototype digital microscope with a motorized stage.
3. One AI engineer having expertise in computer vision and proficiency in Keras.
4. One consultant malarial diagnosis technician who can help in building a training set of images.

Phase 2   In the second phase, we will target the endemic markets. The instrument built in Phase 1 will be replaced with an inexpensive version. Instead of a laptop, an embedded platform such as a cell phone will be used. Target cost of the system: $500–$1000. There will be a need for another round of financing of the order of $1M. A small team of five engineers having expertise in AI, software, embedded systems, and field work will optimize the system. Estimated time for full product launch is 9 months.

Marketing Plan and Pricing Model   The customers will be health clinics and hospitals. An influential paper co-authored by a leading malaria expert will be used to rapidly promote the technology. A pricing model based on pay-per-use will be developed.

Answer of Exercise 33

Consider this a futuristic example which illustrates the power of Bayesian inference and how it performs fusion of evidence. An AI robot has been sent to Mars to prepare it for a human colony. It is scouting a place to build a camp. The Martian Soil S comes in two types: (1) Unsafe (U) for humans and (2) Habitable (H) for humans. The prior probability for the two types is as follows:

$$\Pr(\text{Soil} = \text{Unsafe}) = \Pr(S = U) = 0.01,$$

$$\Pr(\text{Soil} = \text{Habitable}) = \Pr(S = H) = 0.99.$$

The robot can conduct two tests A and B to detect the safety of the soil. Test A is more elaborate and expensive. The true positive rate (TPR) and false positive rate (FPR) of the tests are

$$\text{Test } A : \text{TPR} = 0.99 \text{ and FPR} = 0.01,$$

$$\text{Test } B : \text{TPR} = 0.95 \text{ and FPR} = 0.04,$$

which imply that

$$\Pr(A = P|S = U) = 0.99,$$

$$\Pr(A = P|S = H) = 0.01,$$

$$\Pr(B = P|S = U) = 0.95,$$

$$\Pr(B = P|S = H) = 0.04,$$

where P means the positive result. Suppose the robot performs Test A and gets a positive result that S is unsafe. As per Bayes' rule, $\Pr(S = U|A = P)$ is given by

$$\frac{\Pr(A = P|S = U)\Pr(S = U)}{\Pr(A = P|S = U)\Pr(S = U) + \Pr(A = P|S = H)\Pr(S = H)},$$

which is

$$\frac{0.99 \times 0.01}{0.99 \times 0.01 + 0.01 \times (1 - 0.01)} = 0.5.$$

Therefore, the prior $\Pr(S = U)$ has increased from 0.01 to 0.5. This is a revision of belief on the presentation of evidence of Test A.

The power of Bayesian inference comes next. What if new evidence is presented? The new evidence could be a positive result either by repeating Test A or by the other Test B. Repeating Test A gives us

$$\frac{0.99 \times 0.5}{0.99 \times 0.5 + 0.01 \times (1 - 0.5)} = 0.99,$$

making the robot's belief rise to 0.99. Had it performed the inexpensive Test B, then we have

$$\frac{0.95 \times 0.5}{0.95 \times 0.5 + 0.04 \times (1 - 0.5)} = 0.959$$

and therefore the belief rises to 0.959. The reader can verify that we get the same result if Test B is followed by Test A with both giving positive results.

If only Test B is done once with a positive result, the prior belief changes from 0.01 to 0.193. Again the power of Bayesian evidence fusion is apparent when Test B is done a second time with a positive result, making the belief increase from 0.193 to 0.85.

Answer of Exercise 34

(1)   We refer to [18], in which it is shown that word embeddings inherit the biases present in the text corpus. Consider the gender bias as shown by the word analogy learned by the semantic latent space representations,

Man is to Doctor as Woman is to Nurse.

A classifier is built to classify words as gender definitional or gender neutral. Therefore, "queen" is definitional, whereas "nurse" is neutral. The gender direction is the average vector between the definitional males and the definitional females. To remove the bias, the gender direction is subtracted from the neutral words by projecting their embeddings on the orthogonal subspace. In addition, the definitional pairs such as (king, queen) and (father, mother) are equalized so that all of them are equidistant from the orthogonal subspace.

(2)   There is no correct answer here. Feel comfortable with your opinion, and at the same time, be willing to engage in civil productive discourse with people who don't agree with you. We should be prepared to see both wonderful and undesirable uses of AI in the future. The world may move towards greater disparity unless novel democratic solutions are developed to prevent that from happening.

Answer of Exercise 35

Though there are no consistent definitions of the terms, one can define strong AI as AI that seeks to be as general as human intelligence in all respects. It is also called artificial general intelligence (AGI). Sometimes, consciousness is required to be a part of strong AI; see [162]. Weak AI implements a portion of human intelligence, and it appears to be intelligent at the behavioral level without having true understanding. Narrow AI is AI which solves a constrained, specialized problem. Examples of strong AI are the robotic boy David and Specialists (evolved Mecha) from the Hollywood movie A.I. (2001). An example of weak AI is a household robot with conversational abilities in not very distant future, likely to become a reality in a couple of decades. An example of narrow AI is Siri, Google Assistant, or Tesla's self-driving car.

Appendix B
Lab Exercises and Projects

> The true laboratory is the mind, where
> behind illusions we uncover the laws
> of truth.
>
> Jagadish Chandra Bose

We list a number of assignments and projects which will assist the students hone their practical AI skills. A vast amount of easily accessible code already exists on the Internet. Therefore, often the exercise reduces to finding the right code samples, reading them, putting them together, and customizing the solution.

The students should develop experience in Python programming language and key Python packages of *numpy* and *matplotlib*. We recommend *Keras* and *PyTorch* for AI framework. In addition, *Scikit-learn* and *R* should be explored for classical machine learning. For vision applications, *OpenCV* should be explored. For NLU, *NLTK* offers useful functions. For speech applications, AI software frameworks provide utility functions which facilitate file I/O and creation of spectrograms. The students should learn how to work in *Linux* environment, how to install packages, how to use *Jupyter* notebooks, and how to use source control such as *github*.

B.1 Exercises

1. Write a Python script to perform linear regression on a given set of (x, y) input-output samples, where x is d-dimensional input and y is the real-valued output. Note that there is a direct closed-form solution to linear regression problem. It can be also solved using gradient descent. Implement both approaches.
2. Write a Python script to perform a two-class logistic regression on a given (x, y) input-output samples, where x is d-dimensional and y is the categorical output. Display ROC and PR curves.

3. Implement a single neuron with different activation functions: the ReLU, sigmoid, and hyperbolic tangent functions.
4. Implement a simple feed-forward network for regression using an AI software framework. Perform a single forward pass followed by a backward pass. Note that you have to just call forward and backward functions provided by the AI software.
5. Use different available choices, including Xavier (Glorot) initialization, for initializing the learnable parameters of a neural network. Add a batch normalization layer to a small neural network.
6. Implement the Sobel edge detectors as an example of 2-D convolution.
7. Implement a LeNet-like model for MNIST data. Experiment with Fashion-MNIST data. Explore overfitting by training on an extremely small training data. Explore underfitting by over-simplifying the model. Explore how you will plot ROC and PR curves for multi-class problems.
8. Experiment with data loaders which perform data augmentation such as scaling, resizing, rotation, cropping, and normalization.
9. Experiment with data parallelism to expedite the training. You will need a multi-GPU machine or a multi-GPU cloud computing instance.
10. Implement a character-level recurrent (LSTM) model to predict the next character in an English sentence.
11. Train a cat versus not-cat classifier. Turn it into a sliding window cat detector for larger images.
12. Train a single-layer bidirectional recurrent (LSTM) model on the IMDB movie review sentiment classification dataset. Variable-length input sequences should be truncated or padded to a fixed maximum length. Increase the number of LSTM layers to two.
13. For a linear regression problem, plot the learning curves which show the training and test errors against the size of the training data.
14. Explore OpenAI gym for reinforcement learning. Implement Q-learning for one of the simple examples.
15. Collect a set of pairs of similar images. Collect another set of pairs of dissimilar images. For each pair, measure the similarity using the cosine distance on features extracted from any specified layer of a pre-trained model such as VGG, Inception V3, or AlexNet.
16. Implement a U-Net-like network to segment images. Experiment with MNIST dataset. Use binarized MNIST as the ground truth. Experiment with both cross-entropy and Dice loss functions.
17. Test the performance of the CNN-based face detector and the CNN-based text detector which are provided in OpenCV.

18. Perform text classification on the Newsgroup20 dataset using pre-trained Glove word embeddings. Then, repeat this experiment with pre-trained Word2Vec word embeddings. Variable-length input sequences should be truncated or padded to a fixed maximum length.

19. Convert a sentence into a vector by averaging word embedding vectors over all its words. Implement a method to implement semantic similarity of a pair of sentences using the sentence vectors. Does the scheme generalize to semantic similarity of documents? Repeat the experiment with pre-trained transformer-based models such as BERT.

20. Utter different words and record them. Convert the audio files to spectrograms. Visualize spectrograms for different words.

21. Classify speakers using fast Fourier transform (FFT) and a 1-D Convnet. Repeat the experiment using spectrograms and a 2-D Convnet.

22. Implement a variational autoencoder (VAE) trained on MNIST digits which reconstructs the input. Repeat the same experiment for Fashion-MNIST.

23. For MNIST digits, create synthetic variations using classical image processing techniques such as morphological operations. For example, convert an image to an edge image. For another example, convert an image to a dilated version. Implement an autoencoder to translate a given input image to a given variation.

24. Implement collaborative filtering to recommend movies using a model trained on the MovieLens dataset in which users and movies are embedded in a d-dimensional space and the dot product is used to measure the similarity.

B.2   Exploration

1. See Fig. 4.2. For a toy 2-D problem and a small feed-forward multi-layer network, display the carvings performed by different neurons in terms of convex polytopes and bending lines as a gray-scale image. To each point in a convex polytope, convert the output score of the network to the pixel's greyness value.

2. Read Sect. 4.9.2. Repeatedly draw random samples from an $N \times N$ Gaussian orthogonal ensemble. For each sampled matrix, compute the eigenvalues. Compute its index which is the fraction of the negative eigenvalues. Plot a histogram of the index. Do you ever get all eigenvalues to be positive?

3. Read Sect. 4.9.3. Implement the simplest spin glass, a two-spin Ising model. For a configuration, compute its energy. Implement a greedy algorithm to tweak the configuration to lower its energy. Explore the energy landscape of the model, and note down when you get stuck in a local minima.

4. It has been empirically observed that the Hessian matrix for the loss functions of AI models is singular. Display the distribution of eigenvalues of the Hessian matrix at different epochs during the training of a toy AI model.

5. Using t-SNE, explore low-dimensional visualization of features extracted from a layer of a trained CNN model for MNIST. Do you clearly see ten clusters? Repeat for Fashion-MNIST.

6. For MNIST, visualize and interpret the working of a trained CNN model by displaying the gradient of the output probability scores with respect to the input pixels. Choose an image from the training set. Pick one of the ten classes. Take a gradient ascent step by modifying the input pixels to increase the score of the chosen image class. Keep on iterating till you maximize the score. Repeat the experiment starting with random-looking images. The goal is to fool the network with adversarial images which don't make sense. Interpret your findings.

B.3   Debate and Discussion

1. Organize a moderated debate on the assertion that AI is beneficial to the humanity.
2. Organize a moderated debate on the assertion that human intelligence will be always superior to AI.
3. Visualize a distant future in which AI can do everything what humans can do and there is a surplus economy. Discuss in such a world, how the society can be organized, how people can be motivated, and how economy can be restructured.

Further Reading

My primary goal in writing this book has been to bring an intuitive and comprehensive view of AI to everyone with minimal use of mathematics. The field of AI has grown exponentially in the last decade, and therefore the aim of this compactly written book is to build a high-level, unifying, insightful, bird's eye view of a plethora of techniques and their numerous variations for different applications. I have written this book based on hard work by many scientists and engineers which has led to remarkable, cumulative, community-wide progress. Even though it is not feasible to give credit to everyone individually, I acknowledge the contributions of all in the community which made this book possible.

Here are some tips to students who wish to dive deeper into the field which I have found useful for my own lifelong learning. Reading the classical ML books remains a useful exercise; see [16, 51, 103]. Then there are specialized books in deep learning; see [2, 62] for a comprehensive, detailed coverage of the field. Practitioners should refer to [35, 57] to pick up AI coding skills. For background on reinforcement learning, see [207].

The Internet has become a great resource to learn AI. Excellent courses by Geoffrey Hinton and Andrew Ng provide valuable intuition into difficult topics with the help of simple examples; see [41, 84, 149, 150]. The websites of AI software frameworks [233, 237, 238] have sample code for many problems. Nothing is more informative than looking at the code behind an algorithm.

The ArXiv website has a huge collection of public-access articles in ML and AI; see [227, 228]. It has become the primary resource for me to access and read research papers. With the rise of the public-domain model of publication in which the source code is released as part of sharing of one's work, we are witnessing a paradigm shift towards immediately verifiable, readily reusable, open-access research.

The sky is the limit in learning. Learning never ends.

S. Dube, *An Intuitive Exploration of Artificial Intelligence*,
https://doi.org/10.1007/978-3-030-68624-6

Glossary

Q-**value** Quality of an action for a given state in reinforcement learning as measured by the future expected reward .

activation function A function applied at the output of a neuron to transform it non-linearly.

artificial intelligence An interdisciplinary field which seeks to create intelligent machines. The definition of intelligence is behavior which is at par with or better than that of humans and which involves processes such as learning, adapting, generalization and responsiveness.

attention mechanism Addition of a memory to the encoder in the encoder-decoder architecture of AI models to save the encoder states, which is selectively used by the decoder to augment its input. In other words, the decoder selectively attends to the encoder states during decoding.

attribute learning The same as discriminative machine learning where there is additional semantic interpretation of the output in terms of attributes of the input.

autoencoder A generative model in which the encoder outputs the latent generative parameters for an input. The decoder takes this latent representation as its input and attempts to reconstruct the original input .

backpropagation The chain rule of differentiation implemented for computation graphs .

backward pass The flow of computation from the loss function, through the output, to the input in a computation graph which implements the chain rule of differentiation.

Bayesian inference Application of Bayes' theorem to update beliefs on presentation of new evidence .

© The Author(s), under exclusive license to Springer Nature Switzerland AG 2021
S. Dube, *An Intuitive Exploration of Artificial Intelligence*,
https://doi.org/10.1007/978-3-030-68624-6

bias-variance tradeoff   An AI / ML model can be characterized in terms of its bias and variance, where the former refers to its inherent assumptions and limitations, and the latter refers to its dependence on fluctuations in the training data. Though there is a tradeoff between the two, both can be minimized through careful design of the system and adequate training data.

Boltzmann machine   A generative model which forms a graph of visible nodes and hidden nodes. The observed data is input to the visible nodes and the hidden nodes create its latent representation .

capsule neural network   A deep neural network in which groups of neurons are grouped into capsules which detect parts of an object and output vectors describing the parts, which are then hierarchically combined to detect bigger parts or the whole using an iterative algorithm which optimizes agreement or consensus between capsules .

classification   A classification problem in AI and ML is a type of problem in which the input is categorized. Therefore the output is a categorical variable.

computation graph   A graph in which the nodes perform operations and the results flow through the edges.

connectionist temporal classification   A type of probability scoring output of a neural network used in speech and handwriting recognition with variable-length inputs mapping to the same output .

convex polytope   A polytope is a geometrical object with flat sides. Convexity means that a line segment drawn between any two points belonging to the polytope will lie inside the polytope. Simple examples in 3-D are pyramids, cuboids and parallelepipeds.

convolutional neural network   A deep neural network which uses convolutional layers. A convolutional layer performs the convolution operation, which is a multiply-and-add sliding window operation, on the output of the previous layers .

cross-entropy loss   A loss function for classification problems based on the maximum likelihood estimation principle. Also known as the logistic loss.

deep belief network   A generative model formed by stacking multiple restricted Boltzmann machines or autoencoders .

deep neural network   Generalization of the classical shallow neural network by adding multiple hidden layers. Several variations exist targeted at different applications. A deep neural network has an underlying directed computation graph.

deep reinforcement learning   A specific way to implement reinforcement learning by using a deep neural network to predict the future expected reward of an action for a given state .

discriminative machine learning   Machine learning which takes as input a data sample and outputs values of desired attributes including classification labels. It performs classification or regression or both.

distributed representation   Representing a symbolic entity such as a word, a movie, or a concept as an activation pattern of several neurons. A neuron can contribute to the representation of many entities. This is contrasted with logic-based systems, which treat an entity as a discrete symbol.

embedding   A numerical vector representation of a categorical or discrete entity such as a word, sentence or document in natural language understanding. The concept can be generalized to categorical variables which occur in the context of each other.

encoder-decoder architecture   Generalization of the autoencoder architecture to diverse situations across many application domains in which the encoder builds an internal representation of the input which is decoded by the decoder to produce the output .

end-to-end solution   An approach to build an overall solution by using one integrated AI model which maps the input to the final output. This is contrasted with a component-based design, in which the system is divided into smaller pieces .

expressive power   A measure of the capacity of an AI or ML model to approximate functions. The higher the expressive power, the higher the capacity to fit more complex functions.

feature engineering   A manual process of feature extraction and selection in classical ML based on human expertise.

feed-forward network   A neural network which has a computation graph without cycles or states.

forward pass   The flow of computation from the input to the output in a computation graph.

fully connected layer   A layer in a neural network in which neurons are connected to every neuron of the next layer.

gated recurrent unit   A simplified variant of the long short term memory cell with fewer parameters .

generative adversarial network   A generative model trained through a competitive game with an adversarial discriminative network which evaluates the generated samples .

generative machine learning   Machine learning which can generate data samples of a desired class. The input to a generative model is a set of generative parameters which explicitly or implicitly define the desired properties of the object to be generated.

gradient descent An algorithm which modifies a parameter to minimize a function by an amount given by the negative of the gradient of the function with respect to the parameter and scaled by a learning rate.

hidden layer An intermediate layer in a neural network between the input layer and the output layer.

inference Application of a trained AI or ML model on an unseen data sample.

input layer A pass-through layer of a neural network which receives the input and passes it to the first processing layer.

intelligence The ability to learn, adapt, plan, reason, understand, solve, respond, communicate, build and create.

learning curve A plot of the test error and the training error as a function of the number of training samples. A second kind of learning curve shows dependence on the expressive power of the AI/ML model.

linear model A classical ML model which performs regression or classification using a linear function. For classification, there is an additional step of squishing the output to the [0, 1] range.

linear regression A classical ML model for regression using a linear function.

logistic regression A classical ML model for classification using a linear function followed by the sigmoid or the softmax function.

long short-term memory A variant of recurrent neural networks in which neurons have a gated mechanism to control how memory is retained or forgotten, allowing it learn both long-term and short-term dependencies .

loss function A function to measure the departure of the output of an AI or ML model from the ground truth.

machine learning An interdisciplinary field overlapping computer science, statistics and mathematics which can be viewed as a sub-field within AI and which is applied to solve specific problems through the process of learning from data instead of being explicitly programmed.

manifold A manifold is a set such that the neighborhood of each point is Euclidean. That is, each point has a neighborhood which is homeomorphic to the Euclidean space.

maximum likelihood principle A statistical inference method which estimates the parameters of a model by maximizing the probability of the observed data. The cross-entropy loss, the L^1 loss and the L^2 loss are based on this principle. Bayesian inference under a non-informative prior is equivalent to maximum likelihood estimation .

multilayer perceptron A type of feed-forward network which has multiple layers of perceptrons, a highly specialized binary classifier with a step activation function, studied in the early days of neural network research. Over time, the definition has expanded and its distinction from the feed-forward network has blurred.

natural language understanding   The application of computer science to solve problems on text documents which requires semantic analysis.

object detection   A computer vision task to find all objects of interest and their locations in a given image.

optimization landscape   Plot of the loss function to be minimized as a function of the learnable parameters.

output layer   A processing layer of a neural network whose output is the final output of the entire neural network.

overfitting   Having low training error and high test error due to high expressive power of an AI/ML model and lack of sufficient data.

policy gradient   The gradient of the objective function of the expected rewards with respect to the parameters of the policy network .

rectified linear unit (ReLU)   An activation function defined as $\max\{0, z\}$ which allows only a positive value to pass through.

recurrent neural network   A deep neural network, with memory in its hidden-layer neurons, which processes a sequential input. The memory is called the state of the neuron, which is updated at every time step .

regression   A regression problem in AI and ML is a type of problem in which a numerical attribute of the input is computed. Therefore the output is often a real, continuous number, but it can be an ordinal or cardinal in the more general case.

regularization   Adding a constraint function to an objective function to be optimized. This is the same as constrained optimization in calculus using the Lagrange multiplier method.

reinforcement learning   A sub-field of machine learning in which an agent explores an environment seeking to maximize rewards by selecting actions depending on its current state and observations according to a policy which is learned over time .

restricted Boltzmann machine   A two-layer generative model which forms a bipartitle graph of visible nodes and hidden nodes. The observed data is input to the visible layer and the hidden layer creates its latent representation .

semantic segmentation   Labeling each pixel of an image with the label of the object it belongs to .

softmax function   A generalization of the sigmoid function to multi-class classification. It squishes an arbitrary real-valued vector to a probability vector.

stochastic gradient descent   A variant of gradient descent algorithm in which gradient steps are taken on randomly selected minibatches of the training data.

support vector machine   A classical ML model which performs classification using a maximum-margin classifier. It uses the kernel trick to implicitly map the input feature space to a high-dimensional space, where it linearly separates out the classes by finding the widest separating margin. It can be generalized to regression.

test error   The result of a suitably chosen evaluation metric computed over the test set.

training error   The result of a suitably chosen evaluation metric computed over the training set.

training process   Fitting a machine learning model to the training data with the goal of keeping the generalization error as low as possible.

transformer   A deep neural network which builds a rich, contextual, order-aware representation of tokens in a given input sequence using self-attention layers. It is frequently used in NLU as the encoder in an encoder-decoder architecture .

underfitting   Having high training error due to low expressive power of an AI/ML model.

unsupervised learning   Methods employed to understand the inherent geometry of and to discover patters in unlabeled data. This is contrasted with supervised learning, which needs labeled (annotated) data.

variational autoencoder   A generative model in which an encoder outputs statistical parameters, typically means and variances of Gaussian distributions, of the latent variables of the given input, in contrast to the standard autoencoder, which outputs the values of the latent variables. The decoder randomly chooses a sample from the Gaussian distributions of the latent variables to reconstruct the input .

References

1. Sakshi Agarwal and Lav R. Varshney. *Limits of Deepfake Detection: A Robust Estimation Viewpoint*. 2019. arXiv: 1905.03493 [cs.LG].
2. Charu C. Aggarwal. *Neural Networks and Deep Learning*. Springer, 2018. ISBN: 978-3-319-94462-3.
3. Forest Agostinelli et al. "Solving the Rubik's Cube with Deep Re-inforcement Learning and Search". In: *Nature Machine Intelligence* (2019), pp. 1–8.
4. Martín Arjovsky, Soumith Chintala, and Léon Bottou. "Wasserstein Generative Adversarial Networks". In: *ICML*. 2017.
5. E. Arnold et al. "A Survey on 3D Object Detection Methods for Autonomous Driving Applications". In: *IEEE Transactions on Intelligent Transportation Systems* 20.10 (2019), pp. 3782–3795.
6. Antonio Auffinger, Gérard Ben Arous, and Jiří Černý "Random Matrices and Complexity of Spin Glasses". In: *Communications on Pure and Applied Mathematics* 66.2 (2013), pp. 165–201. DOI: https://doi.org/10.1002/cpa.21422.
7. Vijay Badrinarayanan, Alex Kendall, and Roberto Cipolla. "SegNet: A Deep Convolutional Encoder-Decoder Architecture for Image Segmentation". In: *IEEE Transactions on Pattern Analysis and Machine Intelligence* 39 (2017), pp. 2481–2495.
8. Claudine Badue et al. *Self-Driving Cars: A Survey*. 2019. arXiv: 1901. 04407 [cs.RO].
9. Dzmitry Bahdanau, Kyunghyun Cho, and Yoshua Bengio. *Neural Machine Translation by Jointly Learning to Align and Translate*. 2014. arXiv: 1409.0473 [cs.CL].
10. David Bau et al. "Understanding the Role of Individual Units in a Deep Neural Network". In: *Proceedings of the National Academy of Sciences* (2020). DOI: https://doi.org/10.1073/pnas.1907375117.
11. Erik J. Bekkers et al. "Roto-Translation Covariant Convolutional Networks for Medical Image Analysis". In: *MICCAI*. 2018.
12. Rachel K. E. Bellamy et al. "AI Fairness 360: An Extensible Toolkit for Detecting and Mitigating Algorithmic Bias". In: *IBM J. Res. Dev.* 63 (2019), 4:1–4:15.
13. Emily M. Bender and Batya Friedman. "Data Statements for Natural Language Processing: Toward Mitigating System Bias and Enabling Better Science". In: *Transactions of the Association for Computational Linguistics* 6 (2018), pp. 587–604.
14. Yoshua Bengio et al. "A Neural Probabilistic Language Model". In: *J. Mach. Learn. Res.* 3 (2000), pp. 1137–1155.
15. Yoshua Bengio et al. "Greedy Layer-Wise Training of Deep Networks". In: *NIPS*. 2006.
16. Christopher M. Bishop. *Pattern Recognition and Machine Learning* Springer, 2006. ISBN: 978-0387310732.

© The Author(s), under exclusive license to Springer Nature Switzerland AG 2021
S. Dube, *An Intuitive Exploration of Artificial Intelligence*,
https://doi.org/10.1007/978-3-030-68624-6

17. Flickr Blog. *Introducing Similarity Search at Flickr*. 2017. URL: https://code.flickr.net/2017/03/07/introducing-similarity-search-at-flickr/.
18. Tolga Bolukbasi et al. "Man is to Computer Programmer as Woman is to Homemaker? Debiasing Word Embeddings". In: *Advances in Neural Information Processing Systems 29*. 2016, pp. 4349–4357.
19. Andrew Brock, Jeff Donahue, and Karen Simonyan. "Large Scale GAN Training for High Fidelity Natural Image Synthesis". In: *ICLR*. 2019.
20. Tom B. Brown et al. *Language Models Are Few-Shot Learners*. 2020. arXiv: 2005.14165 [cs.CL].
21. Joy Buolamwini and Timnit Gebru. "Gender Shades: Intersectional Accuracy Disparities in Commercial Gender Classification". In: *FAT*. 2018.
22. Christopher J. C. Burges. *From RankNet to LambdaRank to Lamb-daMART: An Overview*. Tech. rep. Microsoft Research, 2010.
23. Isabel Cachola et al. *TLDR: Extreme Summarization of Scientific Documents*. 2020. arXiv: 2004.15011 [cs.CL].
24. Tim Capes et al. "Siri On-Device Deep Learning-Guided Unit Selection Text-to-Speech System". In: *INTERSPEECH*. 2017.
25. Nicolas Carion et al. *End-to-End Object Detection with Transformers*. 2020. arXiv: 2005.12872 [cs.CV].
26. D. Chalmers. *The Conscious Mind*. Oxford University Press, 1996.
27. William Chan et al. "Listen, Attend and Spell: A Neural Network for Large Vocabulary Conversational Speech Recognition". In: *IEEE International Conference on Acoustics, Speech and Signal Processing (ICASSP)* (2016), pp. 4960–4964.
28. Irene Y. Chen, Fredrik D. Johansson, and David Sontag. "Why is My Classifier Discriminatory?" In: *Proceedings of the 32nd International Conference on Neural Information Processing Systems*. NeurIPS'18. 2018, pp. 3543–3554.
29. Liang-Chieh Chen et al. "DeepLab: Semantic Image Segmentation with Deep Convolutional Nets, Atrous Convolution, and Fully Connected CRFs". In: *IEEE Transactions on Pattern Analysis and Machine Intelligence* 40 (2018), pp. 834–848.
30. Tianqi Chen and Carlos Guestrin. "XGBoost: A Scalable Tree Boosting System". In: *Proceedings of the 22nd ACM SIGKDD International Conference on Knowledge Discovery and Data Mining* (2016).
31. Xi Chen et al. "Variational Lossy Autoencoder". In: *ArXiv* abstract /1611.02731 (2017).
32. Zhe Chen. "Bayesian Filtering: From Kalman Filters to Particle Filters, and Beyond". In: *Statistics* 182 (2003).
33. C. Chiu et al. "State-of-the-Art Speech Recognition with Sequence-to-Sequence Models". In: *IEEE International Conference on Acoustics, Speech and Signal Processing (ICASSP)*. 2018, pp. 4774–4778.
34. Kyunghyun Cho et al. "Learning Phrase Representations using RNN Encoder-Decoder for Statistical Machine Translation". In: *Empirical Methods in Natural Language Processing*. 2014, pp. 1724–1734.
35. Francois Chollet. *Deep Learning with Python*. 1st. USA: Manning Publications Co., 2017. ISBN: 1617294438.
36. François Chollet. "Xception: Deep Learning with Depthwise Separable Convolutions". In: *IEEE Conference on Computer Vision and Pattern Recognition (CVPR)* (2017), pp. 1800–1807.
37. A. Choromanska et al. "The Loss Surfaces of Multilayer Networks". In: *Journal of Machine Learning Research* 38.8 (2015), pp. 192–204.
38. Ö. Çiçek et al. "3D U-Net: Learning Dense Volumetric Segmentation from Sparse Annotation". In: *Medical Image Computing and Computer-Assisted Intervention (MICCAI)*. Vol. 9901. LNCS. 2016, pp. 424–432.
39. Melanie Clapham et al. "Automated Facial Recognition for Wildlife that Lack Unique Markings: A Deep Learning Approach for Brown Bears". In: *Ecology and Evolution* (2020). DOI: https://doi.org/10.1002/ece3.6840.

40. Taco S. Cohen et al. *Gauge Equivariant Convolutional Networks and the Icosahedral CNN*. 2019. arXiv: 1902.04615 [cs.LG].

41. CS231n. *Stanford AI Course*. 2016.

42. Y. Dauphin et al. "Identifying and Attacking the Saddle Point Problem in High-Dimensional Non-convex Optimization". In: *Proceedings of Neural Information Processing Systems*. 2014.

43. Jesse Davis and Mark Goadrich. "The Relationship Between Precision-Recall and ROC Curves". In: *ICML 06*. 2006.

44. David S. Dean and Satya N. Majumdar. "Large Deviations of Extreme Eigenvalues of Random Matrices". In: *Phys. Rev. Lett.* 97 16 (2006), pp. 160–201. DOI: https://doi.org/10.1103/PhysRevLett.97.160201.

45. Jean-Pierre Dedieu and Gregorio Malajovich. "On the Number of Minima of a Random Polynomial". In: *Journal of Complexity* 24.2 (2008), pp. 89–108. issn: 0885-064X. DOI: https://doi.org/10.1016/j.jco.2007.09.003.

46. Michäel Defferrard, Xavier Bresson, and Pierre Vandergheynst. "Convolutional Neural Networks on Graphs with Fast Localized Spectral Filtering". In: *NIPS*. 2016.

47. J. Deng et al. "ImageNet: A Large-Scale Hierarchical Image Database". In: *IEEE Conference on Computer Vision and Pattern Recognition*. 2009, pp. 248–255.

48. Jacob Devlin et al. "BERT: Pre-training of Deep Bidirectional Transformers for Language Understanding". In: *ArXiv* abstract /1810.04805 (2019).

49. Jeff Donahue and Karen Simonyan. "Large Scale Adversarial Representation Learning". In: *NeurIPS*. 2019.

50. Simant Dube. *High Dimensional Spaces, Deep Learning and Adversarial Examples*. 2018. arXiv: 1801.00634 [cs.CV].

51. R. Duda, P. Hart, and D. Stork. *Pattern classification*. John Wiley and Son, 2001.

52. Shai Fine, Yoram Singer, and Naftali Tishby. "The Hierarchical Hidden Markov Model: Analysis and Applications". In: *Machine Learning* 32 (1998), pp. 41–62.

53. Tim Fountaine, Brian McCarthy, and Tamim Saleh. "Building the AI-Powered Organization". In: *Harvard Business Review* (July 2019), pp. 62–73.

54. Jiyang Guo et al. *VectorNet: Encoding HD Maps and Agent Dynamics from Vectorized Representation*. 2020. arXiv: 2005.04259 [cs.CV].

55. Timur Garipov et al. "Loss Surfaces, Mode Connectivity, and Fast Ensembling of DNNs". In: *NeurIPS*. 2018.

56. Leon A. Gatys, Alexander S. Ecker, and Matthias Bethge. *A Neural Algorithm of Artistic Style*. 2015. arXiv: 1508.06576 [cs.CV].

57. Aurélien Géron. *Hands-On Machine Learning with Scikit-Learn and TensorFlow: Concepts, Tools, and Techniques to Build Intelligent Systems*. O'Reilly Media, 2017. ISBN: 978-1491962299.

58. Xavier Glorot and Yoshua Bengio. "Understanding the Difficulty of Training Deep Feedforward Neural Networks". In: *AISTATS*. 2010.

59. Xavier Glorot, Antoine Bordes, and Yoshua Bengio. "Deep Sparse Rectifier Neural Networks". In: *AISTATS*. 2011.

60. Ian J. Goodfellow and Oriol Vinyals. "Qualitatively Characterizing Neural Network Optimization Problems". In: *ICLR*. 2015.

61. Ian J. Goodfellow et al. "Generative Adversarial Nets". In: *Neural Information Processing Systems (NeurIPS)*. 2014.

62. Ian Goodfellow, Yoshua Bengio, and Aaron Courville. *Deep Learning*. The MIT Press, 2017. ISBN: 978-0262035613.

63. Priya Goyal et al. *Accurate, Large Minibatch SGD: Training ImageNet in 1 Hour*. 2018. arXiv: 1706.02677 [cs.CV].

64. Benjamin Graham and Laurens van der Maaten. *Submanifold Sparse Convolutional Networks*. 2017. arXiv: 1706.01307 [cs.NE].

65. Alex Graves and Navdeep Jaitly. "Towards End-to-End Speech Recognition with Recurrent Neural Networks". In: *Proceedings of the 31st International Conference on International Conference on Machine Learning - Volume 32*. 2014.

66. Alex Graves, Greg Wayne, and Ivo Danihelka. *Neural Turing Machines*. 2014. arXiv: 1410.5401 [cs.NE].

67. Alex Graves et al. "A Novel Connectionist System for Unconstrained Handwriting Recognition". In: *IEEE Transactions on Pattern Analysis and Machine Intelligence* 31 (2009), pp. 855–868.

68. Alex Graves et al. "Connectionist Temporal Classification: Labelling Unsegmented Sequence Data with Recurrent Neural Networks". In: *ICML '06*. 2006.

69. Alex Graves et al. "Hybrid Computing Using a Neural Network with Dynamic External Memory". In: *Nature* 538 (2016), pp. 471–476.

70. Sorin Grigorescu et al. "A Survey of Deep Learning Techniques for Autonomous Driving". In: *J. Field Robotics* 37 (2020), pp. 362–386.

71. Varun Gulshan et al. "Development and Validation of a Deep Learning Algorithm for Detection of Diabetic Retinopathy in Retinal Fundus Photographs." In: *JAMA* 316 22 (2016), pp. 2402–2410.

72. Yulan Guo et al. "Deep Learning for 3D Point Clouds: A Survey". In: *IEEE Transactions on Pattern Analysis and Machine Intelligence* (2019).

73. Stuart R. Hameroff and Roger Penrose. "Consciousness in the Universe: a Review of the 'Orch OR' Theory." In: *Physics of Life Reviews* 11 1 (2011), pp. 39–78.

74. Awni Y. Hannun et al. "Cardiologist-Level Arrhythmia Detection and Classification in Ambulatory Electrocardiograms Using a Deep Neural Network". In: *Nature Medicine* 25 (2019), pp. 65–69.

75. Yuval N. Harari. *Sapiens: a Brief History of Humankind*. Harper, New York, 2015.

76. Daniel A. Hashimoto et al. "Artificial Intelligence in Surgery: Promises and Perils". In: *Annals of Surgery* 268 (2018), pp. 70–76.

77. Hado van Hasselt, Arthur Guez, and David Silver. "Deep Reinforcement Learning with Double Q-Learning". In: *Proceedings of the Thirtieth AAAI Conference on Artificial Intelligence*. AAAI, 2016, pp. 2094–2100.

78. Kaiming He et al. "Mask R-CNN". In: *IEEE International Conference on Computer Vision (ICCV)* (2017), pp. 2980–2988.

79. K. He et al. "Deep Residual Learning for Image Recognition". In: *IEEE Conference on Computer Vision and Pattern Recognition (CVPR)*. 2016, pp. 770–778.

80. Xiangnan He et al. "Neural Collaborative Filtering". In: *Proceedings of the 26th International Conference on World Wide Web* (2017).

81. Martin Heusel et al. "GANs Trained by a Two Time-Scale Update Rule Converge to a Local Nash Equilibrium". In: *Proceedings of the 31st International Conference on Neural Information Processing Systems*. 2017, pp. 6629–6640.

82. Irina Higgins et al. "beta-VAE: Learning Basic Visual Concepts with a Constrained Variational Framework". In: *ICLR*. 2017.

83. G. E. Hinton, J. McClelland, and D. Rumelhart. "Distributed Representations". In: ed. by D. E. Rumelhart and J. L. McClelland. Vol. 1. Parallel Distributed Processing: Explorations in the Microstructure of Cognition. MIT Press, Cambridge, 1986, pp. 77–109.

84. Geoffrey Hinton. *Neural Networks for Machine Learning. Coursera lectures*. 2013.

85. Geoffrey E. Hinton. "A Practical Guide to Training Restricted Boltzmann Machines". In: *Neural Networks: Tricks of the Trade*. 2012.

86. Geoffrey E. Hinton. "Mapping Part-Whole Hierarchies into Connectionist Networks". In: *Artif. Intell.* 46 (1990), pp. 47–75.

87. Geoffrey E. Hinton, James L. McClelland, and David E. Rumelhart. "Distributed Representations". In: *The Philosophy of Artificial Intelligence*. 1990.

88. Geoffrey E. Hinton, Simon Osindero, and Yee Whye Teh. "A Fast Learning Algorithm for Deep Belief Nets". In: *Neural Computation* 18 (2006), pp. 1527–1554.

89. Geoffrey E. Hinton, Sara Sabour, and Nicholas Frosst. "Matrix Capsules with EM Routing". In: *ICLR*. 2018.

90. Geoffrey E. Hinton and Ruslan Salakhutdinov. "Reducing the Dimensionality of Data with Neural Networks." In: *Science* 313 5786 (2006), pp. 504–507.

91. Geoffrey E. Hinton et al. "Deep Neural Networks for Acoustic Modeling in Speech Recognition: The Shared Views of Four Research Groups". In: *IEEE Signal Processing Magazine* 29 (2012), pp. 82–97.

92. Geoffrey Hinton, Oriol Vinyals, and Jeff Dean. "Distilling the Knowledge in a Neural Network". In: *NIPS Deep Learning and Representation Learning Workshop*. 2015.

93. S. Hochreiter and J. Schmidhuber. "Long short-term memory". In: *Neural Computation* 9.8 (1997), pp. 1735–1780.

94. Elad Hoffer and Nir Ailon. "Deep Metric Learning Using Triplet Network". In: *SIMBAD*. 2015.

95. J. Hu, L. Shen, and G. Sun. "Squeeze-and-Excitation Networks". In: *IEEE/CVF Conference on Computer Vision and Pattern Recognition*. 2018, pp. 7132–7141.

96. Gao Huang, Zhuang Liu, and Kilian Q. Weinberger. "Densely Connected Convolutional Networks". In: *IEEE Conference on Computer Vision and Pattern Recognition (CVPR)* (2017), pp. 2261–2269.

97. D. H. Hubel and T. N. Wiesel. "Receptive Fields, Binocular Interaction, and Functional Architecture in the Cat's Visual Cortex". In: *Journal of Physiology* 160 (1962), pp. 106–154.

98. Forrest N. Iandola et al. *SqueezeNet: AlexNet-Level Accuracy with 50× Fewer Parameters and < 0.5MB Model Size.* 2016. arXiv: 1602.07360 [cs.CV].

99. Marco Iansiti and Karim R. Lakhani. *Competing in the Age of AI: Strategy and Leadership when Algorithms and Networks Run the World.* Harvard Business Review Press, 2020.

100. Sergey Ioffe and Christian Szegedy. "Batch Normalization: Accelerating Deep Network Training by Reducing Internal Covariate Shift". In: *ICML*. 2015.

101. Phillip Isola et al. "Image-to-Image Translation with Conditional Adversarial Networks". In: *IEEE Conference on Computer Vision and Pattern Recognition (CVPR)* (2017), pp. 5967–5976.

102. Max Jaderberg et al. "Spatial Transformer Networks". In: *Advances in Neural Information Processing Systems*. 2015, pp. 2017–2025.

103. Gareth James et al. *An Introduction to Statistical Learning: With Applications in R.* Springer Publishing Company, Incorporated, 2014. ISBN: 1461471370.

104. S. Jégou et al. "The One Hundred Layers Tiramisu: Fully Convolutional DenseNets for Semantic Segmentation". In: *2017 IEEE Conference on Computer Vision and Pattern Recognition Workshops* (2017), pp. 1175–1183.

105. Eun Seo Jo and Timnit Gebru. "Lessons from Archives: Strategies for Collecting Sociocultural Data in Machine Learning". In: *Proceedings of the 2020 Conference on Fairness, Accountability, and Transparency* (2020).

106. Vinay Joshi et al. "Accurate Deep Neural Network Inference Using Computational Phase-Change Memory". In: *Nature Communications* 11 (2020).

107. Yannis Kalantidis and Yannis Avrithis. "Locally Optimized Product Quantization for Approximate Nearest Neighbor Search". In: *IEEE CVPR* (2014), pp. 2329–2336.

108. Alexandre Kaspar et al. "Neural Inverse Knitting: From Images to Manufacturing Instructions". In: *ICML*. 2019.

109. Angelos Katharopoulos et al. *Transformers are RNNs: Fast Autoregressive Transformers with Linear Attention.* 2020. arXiv: 2006.16236 [cs.LG].

110. Daniel J. Kersten, Pascal Mamassian, and Alan L. Yuille. "Object Perception as Bayesian Inference." In: *Annual Review of Psychology* 55 (2004), pp. 271–304.

111. Seung Wook Kim et al. "Learning to Simulate Dynamic Environments With GameGAN". In: *The IEEE/CVF Conference on Computer Vision and Pattern Recognition (CVPR)*. May 2020.

112. Diederik P. Kingma and Max Welling. "Auto-Encoding Variational Bayes". In: *CoRR* abstract /1312.6114 (2014).

113. Scott Kirkpatrick and Bart Selman. "Critical Behavior in the Satisfiability of Random Boolean Expressions". In: *Science* 264 (1994), pp. 1297–1301.

114. Christof Koch. "Proust Among the Machines". In: *Scientific American* (Dec. 2019), pp. 46–49.

115. Yehuda Koren, Robert M. Bell, and Chris Volinsky. "Matrix Factorization Techniques for Recommender Systems". In: *Computer* 42 (2009).

116. A. Krizhevsky, I. Sutskever, and G. Hinton. "ImageNet Classification with Deep Convolutional Neural Networks". In: *Proceedings of Neural Information Processing Systems*. 2012.

117. Alex Krizhevsky. *Learning Multiple Layers of Features from Tiny Images*. Tech. rep. University of Toronto, 2009.

118. S. Kuutti et al. "A Survey of Deep Learning Applications to Autonomous Vehicle Control". In: *IEEE Transactions on Intelligent Transportation Systems* (2020), pp. 1–22.

119. Brenden M. Lake, Ruslan Salakhutdinov, and Joshua B. Tenenbaum. "Human-Level Concept Learning Through Probabilistic Program Induction". In: *Science* 350 (2015), pp. 1332–1338.

120. Guillaume Lample and Francois Charton. "Deep Learning for Symbolic Mathematics". In: *ICLR*. 2020.

121. Yann LeCun, Y. Bengio, and Geoffrey Hinton. "Deep Learning". In: *Nature* 521 (May 2015), pp. 436–44. DOI: 10.1038/nature14539.

122. Yann LeCun et al. "Efficient BackProp". In: *Neural Networks: Tricks of the Trade*. 1998.

123. Yann LeCun et al. "Handwritten Digit Recognition with a Back-Propagation Network". In: *NIPS*. 1989.

124. Y. LeCun et al. "Gradient-Based Learning Applied to Document Recognition". In: *Proceedings of the IEEE* 86.11 (1998), pp. 2278–2324.

125. Jinhyuk Lee et al. "BioBERT: a Pre-trained Biomedical Language Representation Model for Biomedical Text Mining". In: *Bioinformatics* (2020).

126. John Lee. *Introduction to Topological Manifolds*. Graduate Texts in Mathematics. Springer-Verlag New York, 2011. ISBN: 978-1-4419-7940-7.

127. An Open Letter. *Research Priorities for Robust and Beneficial Artificial Intelligence*. 2016. URL: https://futureoflife.org/ai-open-letter/.

128. Chunyuan Li et al. "Measuring the Intrinsic Dimension of Objective Landscapes". In: *ICLR*. 2018.

129. Hao Li et al. "Visualizing the Loss Landscape of Neural Nets". In: *Neural Information Processing Systems*. 2018.

130. Yi Li et al. "Fully Convolutional Instance-Aware Semantic Segmentation". In: *IEEE Conference on Computer Vision and Pattern Recognition (CVPR)* (2017), pp. 4438–4446.

131. Tsung-Yi Lin et al. "Feature Pyramid Networks for Object Detection". In: *IEEE Conference on Computer Vision and Pattern Recognition (CVPR)* (2017), pp. 936–944.

132. Tsung-Yi Lin et al. "Focal Loss for Dense Object Detection". In: *IEEE Transactions on Pattern Analysis and Machine Intelligence* 42 (2020), pp. 318–327.

133. Tsung-Yi Lin et al. *Microsoft COCO: Common Objects in Context*. 2014. arXiv: 1405.0312 [cs.CV].

134. Yujun Lin et al. *Deep Gradient Compression: Reducing the Communication Bandwidth for Distributed Training*. 2017. arXiv: 1712.01887[cs.CV].

135. Wei Liu et al. "SSD: Single Shot MultiBox Detector". In: *ECCV*. 2016.

136. Andrew Lucas. "Ising Formulations of Many NP Problems". In: *Frontiers in Physics* 2 (2014), p. 5. DOI: 10.3389/fphy.2014.00005.

137. Thang Luong, Hieu Pham, and Christopher D. Manning. "Effective Approaches to Attention-Based Neural Machine Translation". In: *Proc. of EMNLP*. 2015.

138. Roland Maas et al. "Anchored Speech Detection". In: *Interspeech*. Sept. 2016, pp. 2963–2967 DOI: 10.21437/Interspeech.2016-1346.

139. Gary Marcus. *Deep Learning: A Critical Appraisal*. 2018. arXiv: 1801.00631[cs.AI].

140. Jonathan Masci et al. "Geodesic Convolutional Neural Networks on Riemannian Manifolds". In: *IEEE International Conference on Computer Vision Workshop (ICCVW)* (2015), pp. 832–840.

141. Cade Metz. "Moguls and Killer Robots". In: *New York Times* (June 10, 2018).

142. Tomas Mikolov et al. *Distributed Representations of Words and Phrases and Their Compositionality*. 2013. arXiv: 1310.4546 [cs.CL].

143. Fausto Milletari, Nassir Navab, and Seyed-Ahmad Ahmadi. "V-Net: Fully Convolutional Neural Networks for Volumetric Medical Image Segmentation". In: *Fourth International Conference on 3D Vision (3DV)* (2016), pp. 565–571.

144. Margaret Mitchell et al. "Model Cards for Model Reporting". In: *Proceedings of the Conference on Fairness, Accountability, and Transparency* (2019).

145. Guido F Montufar et al. "On the Number of Linear Regions of Deep Neural Networks". In: *Advances in Neural Information Processing Systems 27*. Curran Associates, Inc., 2014, pp. 2924–2932.

146. Jonas Mueller and Aditya Thyagarajan. "Siamese Recurrent Architectures for Learning Sentence Similarity". In: *AAAI*. 2016.

147. David Mumford and Agnès Desolneux. *Pattern Theory: The Stochastic Analysis of Real-World Signals*. CRC Press, Jan. 2010.

148. Assaf Neuberger et al. "Image Based Virtual Try-On Network From Unpaired Data". In: *The IEEE/CVF Conference on Computer Vision and Pattern Recognition (CVPR)*. May 2020.

149. Andrew Ng. *Deep Learning Lectures*. 2018. URL: https://www.deeplearning.ai/.

150. Andrew Ng. *Machine Learning. Coursera Lectures*. 2016. URL: https://www.coursera.org/.

151. Anh Nguyen, Jason Yosinski, and Jeff Clune. "Deep Neural Networks are Easily Fooled: High Confidence Predictions for Unrecognizable Images". In: *IEEE Conference on Computer Vision and Pattern Recognition (CVPR)* (2015), pp. 427–436.

152. Thanh Thi Nguyen et al. *Deep Learning for Deepfakes Creation and Detection*. 2019. arXiv: 1909.11573 [cs.CV].

153. Ilkay Oksuz et al. "Deep Learning Based Detection and Correction of Cardiac MR Motion Artefacts During Reconstruction for High-Quality Segmentation". In: *IEEE Transactions on Medical Imaging* (July 2020).

154. B. A. Olshausen and D. J. Field. "Emergence of Simple-Cell Receptive Field Properties by Learning a Sparse Code for Natural Images". In: *Nature* 381 (1996), pp. 607–609.

155. Äaron van den Oord, Nal Kalchbrenner, and Koray Kavukcuoglu. "Pixel Recurrent Neural Networks". In: *ICML*. 2016.

156. Aaron van den Oord et al. *WaveNet: A Generative Model for Raw Audio*. 2016. arXiv: 1609.03499 [cs.SD].

157. Shuyang Shaun Pan et al. "Cosmological Parameter Estimation from Large-Scale Structure Deep Learning". In: *arXiv: Cosmology and Non-galactic Astrophysics* (2019).

158. Austin Peel et al. "Distinguishing Standard and Modified Gravity Cosmologies with Machine Learning". In: *arXiv: Cosmology and Non-galactic Astrophysics* (2018).

159. Jeffrey Pennington, Richard Socher, and Christopher D. Manning. "Glove: Global Vectors for Word Representation". In: *Empirical Methods in Natural Language Processing*. 2014.

160. Roger Penrose. *The Emperor's New Mind: Concerning Computers, Minds, and the Laws of Physics*. USA: Oxford University Press, Inc., 1989. ISBN: 0198519737.

161. Vitali Petsiuk, Abir Das, and Kate Saenko. "RISE: Randomized Input Sampling for Explanation of Black-box Models". In: *BMVC*. 2018.

162. Stanford Encyclopedia of Philosophy. *Artificial Intelligence*. 2018. URL: https://plato.stanford.edu/entries/artificial-intelligence/.

163. Stanford Encyclopedia of Philosophy. *Panpsychism*. 2017. URL: https://plato.stanford.edu/entries/panpsychism/.

164. Stanford Encyclopedia of Philosophy. *Qualia*. 2017. URL: https://plato.stanford.edu/entries/qualia/.

165. Wei Ping et al. "Deep Voice 3: Scaling Text-to-Speech with Convolutional Sequence Learning". In: *arXiv: Sound* (2018).

166. Joan Puigcerver. "Are Multidimensional Recurrent Layers Really Necessary for Handwritten Text Recognition?" In: *14th IAPR International Conference on Document Analysis and Recognition (ICDAR)* 01 (2017), pp. 67–72.

167. Charles Ruizhongtai Qi et al. "PointNet++: Deep Hierarchical Feature Learning on Point Sets in a Metric Space". In: *NIPS*. 2017.

168. Charles Ruizhongtai Qi et al. "PointNet: Deep Learning on Point Sets for 3D Classification and Segmentation". In: *IEEE Conference on Computer Vision and Pattern Recognition (CVPR)* (2017), pp. 77–85.

169. Charles Ruizhongtai Qi et al. "Volumetric and Multi-view CNNs for Object Classification on 3D Data". In: *IEEE Conference on Computer Vision and Pattern Recognition (CVPR)* (2016), pp. 5648–5656.

170. Colin Raffel et al. *Exploring the Limits of Transfer Learning with a Unified Text-to-Text Transformer*. 2019. arXiv: 1910.10683 [cs.LG].

171. Maithra Raghu et al. "On the Expressive Power of Deep Neural Networks". In: *Proceedings of the 34th International Conference on Machine Learning*. Vol. 70. Proceedings of Machine Learning Research. 2017, pp. 2847–2854.

172. Aravind Rajeswaran et al. "Meta-Learning with Implicit Gradients". In: *Advances in Neural Information Processing Systems 32*. 2019, pp. 113–124.

173. Siamak Ravanbakhsh et al. *Estimating Cosmological Parameters from the Dark Matter Distribution*. 2017. arXiv: 1711.02033 [astro-ph.CO].

174. Esteban Real et al. "Regularized Evolution for Image Classifier Architecture Search". In: *Proceedings of the AAAI Conference on Artificial Intelligence*. 2019.

175. Joseph Redmon and Ali Farhadi. *YOLOv3: An Incremental Improvement*. 2018. arXiv: 1804.02767 [cs.CV].

176. J. Redmon et al. "You Only Look Once: Unified, Real-Time Object Detection". In: *IEEE Conference on Computer Vision and Pattern Recognition (CVPR)*. 2016, pp. 779–788.

177. Emily Reif et al. "Visualizing and Measuring the Geometry of BERT". In: *Advances in Neural Information Processing Systems*. Vol. 32. 2019, pp. 8594–8603.

178. Shaoqing Ren et al. "Faster R-CNN: Towards Real-Time Object Detection with Region Proposal Networks". In: *IEEE Transactions on Pattern Analysis and Machine Intelligence* 39 (2015), pp. 1137–1149.

179. Marco Tulio Ribeiro, Sameer Singh, and Carlos Guestrin. "Why Should I Trust You?: Explaining the Predictions of Any Classifier". In: *Proceedings of the 22nd ACM SIGKDD International Conference on Knowledge Discovery and Data Mining* (2016).

180. Olaf Ronneberger, Philipp Fischer, and Thomas Brox. *U-Net: Convolutional Networks for Biomedical Image Segmentation*. 2015. arXiv: 1505.04597 [cs.CV].

181. Mihaela Rosca et al. *Variational Approaches for Auto-Encoding Generative Adversarial Networks*. 2017. arXiv: 1706.04987 [stat.ML].

182. D. E. Rumelhart, G. E. Hinton, and R. J. Willams. "Learning Representations by Back-Propagating Errors". In: *Nature* 323 (1986), pp. 533–536.

183. Sara Sabour, Nicholas Frosst, and Geoffrey E Hinton. *Dynamic Routing Between Capsules*. 2017. arXiv: 1710.09829 [cs.CV].

184. Tim Salimans et al. *PixelCNN++: Improving the PixelCNN with Discretized Logistic Mixture Likelihood and Other Modifications*. 2017. arXiv: 1701.05517 [cs.LG].

185. Artsiom Sanakoyeu et al. "Transferring Dense Pose to Proximal Animal Classes". In: *CVPR*. 2020.

186. Mark Sandler et al. "MobileNetV2: Inverted Residuals and Linear Bottlenecks". In: *IEEE/CVF Conference on Computer Vision and Pattern Recognition* (2018), pp. 4510–4520.

187. F. Schroff, D. Kalenichenko, and J. Philbin. "FaceNet: A unified embedding for face recognition and clustering". In: *IEEE Conference on Computer Vision and Pattern Recognition (CVPR)* (2015), pp. 815–823.

188. Mark P. Sendak et al. "Real-World Integration of a Sepsis Deep Learning Technology into Routine Clinical Care". In: *JMIR Medical Informatics* (2019).

189. Andrew Senior et al. "Improved Protein Structure Prediction Using Potentials from Deep Learning". In: *Nature* 577 (2020), pp. 706–710.

190. Pierre Sermanet et al. "OverFeat: Integrated Recognition, Localization and Detection using Convolutional Networks". In: *ICLR*. 2014.

191. Shikhar Sharma, Ryan Kiros, and Ruslan Salakhutdinov. *Action Recognition Using Visual Attention*. 2015. arXiv: 1511.04119 [cs.LG].

192. Noam Shazeer et al. "Outrageously Large Neural Networks: The Sparsely-Gated Mixture-of-Experts Layer". In: *ICLR*. 2017.
193. Evan Shelhamer, Jonathan Long, and Trevor Darrell. "Fully Convolutional Networks for Semantic Segmentation". In: *IEEE Transactions on Pattern Analysis and Machine Intelligence* 39 (2017), pp. 640–651.
194. C. Shen et al. "Operating a Treatment Planning System using a Deep-Reinforcement-Learning based Virtual Treatment Planner for Prostate Cancer Intensity-Modulated Radiation Therapy Treatment Planning." In: *Medical physics* (2020).
195. Jonathan Shen et al. "Natural TTS Synthesis by Conditioning Wavenet on MEL Spectrogram Predictions". In: *IEEE International Conference on Acoustics, Speech and Signal Processing (ICASSP)* (2018), pp. 4779–4783.
196. David Silver et al. "A General Reinforcement Learning Algorithm that Masters Chess, Shogi, and Go Through Self-Play". In: *Science* 362 (2018), pp. 1140–1144.
197. David Silver et al. "Mastering the Game of Go with Deep Neural Networks and Tree Search". In: *Nature* 529 (2016), pp. 484–489.
198. David Silver et al. "Mastering the Game of Go Without Human Knowledge". In: *Nature* 550 (2017), pp. 354–359.
199. Karen Simonyan and Andrew Zisserman. "Two-Stream Convolutional Networks for Action Recognition in Videos". In: *NIPS*. 2014.
200. Karen Simonyan and Andrew Zisserman. "Very Deep Convolutional Networks for Large-Scale Image Recognition". In: *3rd International Conference on Learning Representations*. 2015.
201. Vincent Sitzmann et al. *Implicit Neural Representations with Periodic Activation Functions*. 2020. arXiv: 2006.09661 [cs.CV].
202. Rupesh Kumar Srivastava, Klaus Greff, and Jürgen Schmidhuber. *Highway Networks*. 2015. arXiv: 1505.00387 [cs.LG].
203. Trevor Standley et al. "Which Tasks Should Be Learned Together in Multi-task Learning?" In: *ICML*. 2020.
204. Richard P. Stanley. "An Introduction to Hyperplane Arrangements". In: *IAS/Park City Mathematics Series*. 2006.
205. Russell J. Stewart, Mykhaylo Andriluka, and Andrew Y. Ng. "End-to-End People Detection in Crowded Scenes". In: *IEEE Conference on Computer Vision and Pattern Recognition (CVPR)* (2016), pp. 2325–2333.
206. Ilya Sutskever, Oriol Vinyals, and Quoc V Le. "Sequence to Sequence Learning with Neural Networks". In: *Advances in Neural Information Processing Systems 27*. 2014, pp. 3104–3112.
207. Richard S. Sutton and Andrew G. Barto. *Reinforcement Learning: An Introduction*. Second. The MIT Press, 2018.
208. Christian Szegedy et al. "Going Deeper with Convolutions". In: *IEEE Conference on Computer Vision and Pattern Recognition* (2015), pp. 1–9.
209. Christian Szegedy et al. "Inception-v4, Inception-ResNet and the Impact of Residual Connections on Learning". In: *AAAI*. 2017.
210. Christian Szegedy et al. *Intriguing Properties of Neural Networks*. 2013. arXiv: 1312.6199 [cs.CV].
211. C. Szegedy et al. "Rethinking the Inception Architecture for Computer Vision". In: *IEEE Conference on Computer Vision and Pattern Recognition (CVPR)*. 2016, pp. 2818–2826.
212. Mingxing Tan and Quoc Le. "EfficientNet: Rethinking Model Scaling for Convolutional Neural Networks". In: *Proceedings of the 36th International Conference on Machine Learning*. 2019, pp. 6105–6114.
213. Justus Thies et al. "Face2Face: Real-Time Face Capture and Reenactment of RGB Videos". In: *Commun. ACM* 62 (2018), pp. 96–104.
214. Eric Topol. *Deep Medicine: How Artificial Intelligence Can Make Healthcare Human Again*. 1st. USA: Basic Books, Inc., 2019. ISBN: 1541644638.
215. H. Tseng et al. "Deep Reinforcement Learning for Automated Radiation Adaptation in Lung Cancer". In: *Medical Physics* 44 (2017), pp. 6690–6705.

216. Silviu-Marian Udrescu and Max Tegmark. "AI Feynman: A Physics-Inspired Method for Symbolic Regression". In: *Science Advances* 6 (2020).
217. Benigno Uria et al. *Neural Autoregressive Distribution Estimation*. 2016. arXiv: 1605.02226 [cs.LG].
218. Ashish Vaswani et al. *Attention Is All You Need*. 2017. arXiv: 1706. 03762 [cs.CL].
219. Pascal Vincent et al. "Stacked Denoising Autoencoders: Learning Useful Representations in a Deep Network with a Local Denoising Criterion". In: *J. Mach. Learn. Res.* 11 (2010), pp. 3371–3408.
220. Boyu Wang et al. "Distribution Matching for Crowd Counting". In: *Advances in Neural Information Processing Systems*. 2020.
221. Chen Wang et al. "6-PACK: Category-Level 6D Pose Tracker with Anchor-Based Keypoints". In: *ICRA*. 2020.
222. Chunhao Wang et al. "Artificial Intelligence in Radiotherapy Treatment Planning: Present and Future". In: *Technology in Cancer Research and Treatment* 18 (2019).
223. Mei Wang and Weihong Deng. *Deep Face Recognition: A Survey*. 2018. arXiv: 1804.06655 [cs.CV].
224. Yu Emma Wang, Gu-Yeon Wei, and David Brooks. *Benchmarking TPU, GPU, and CPU Platforms for Deep Learning*. 2019. arXiv: 1907.10701 [cs.LG].
225. Yuxuan Wang et al. "Tacotron: Towards End-to-End Speech Synthesis". In: *INTERSPEECH*. 2017.
226. Y. Wang et al. "End-to-End Anchored Speech Recognition". In: *IEEE International Conference on Acoustics, Speech and Signal Processing (ICASSP)*. 2019, pp. 7090–7094.
227. ArXiv Website. *A free distribution service and an open-access archive of scholarly articles*. URL: https://arxiv.org/.
228. ArXiv Sanity Website. *Arxiv Sanity Preserver*. URL: http://www.arxiv-sanity.com/.
229. cuDNN Website. *Nvidia Deep Neural Network library*. URL: https://developer.nvidia.com/cudnn.
230. GCP Website. *Google Cloud: Cloud Computing Services*. URL: https://cloud.google.com.
231. GitHub Website. *The world's leading software development platform*. URL: https://github.com/.
232. ImageNet Website. *Large Scale Visual Recognition Challenge*. URL: http://www.image-net.org/challenges/LSVRC/.
233. Keras Website. *The Python deep learning API*. URL: https://keras.io/.
234. MLPerf Website. *Fair and useful benchmarks for measuring training and inference performance of ML hardware, software, and services*. URL: https://mlperf.org/.
235. MONAI Website. *Medical Open Network for AI*. URL: https://monai.io/.
236. OpenVINO Website. *Toolkit to optimize deep learning on Intel platforms*. URL: https://docs.openvinotoolkit.org/.
237. PyTorch Website. *From research to production: an open source machine learning framework*. URL: https://pytorch.org/.
238. TensorFlow Website. *An end-to-end open source machine learning platform*. URL: https://www.tensorflow.org/.
239. Sarah Myers West. "Data Capitalism: Redefining the Logics of Surveillance and Privacy". In: *Business & Society* 58 (2019), pp. 20–41.
240. S.M. West, M. Whittaker, and K. Crawford. *Discriminating Systems: Gender, Race and Power in AI*. 2019. URL: https://ainowinstitute.org/discriminatingsystems.html.
241. Jason Weston, Sumit Chopra, and Antoine Bordes. *Memory Networks*. 2014. arXiv: 1410.3916 [cs.AI].
242. H. James R. Wilson et al. *Artificial Intelligence: The Insights You Need from Harvard Business Review*. Harvard Business Review Press, 2019.
243. Marysia Winkels and Taco S. Cohen. *3D G-CNNs for Pulmonary Nodule Detection*. 2018. arXiv: 1804.04656 [cs.LG].
244. Sam Wiseman and Alexander M. Rush. "Sequence-to-Sequence Learning as Beam-Search Optimization". In: *EMNLP*. 2016.

245. Yonghui Wu et al. *Google's Neural Machine Translation System: Bridging the Gap Between Human and Machine Translation*. 2016. arXiv: 1609.08144 [cs.CL].

246. Saining Xie et al. *Aggregated Residual Transformations for Deep Neural Networks*. 2016. arXiv: 1611.05431 [cs.CV].

247. Kelvin Xu et al. "Show, Attend and Tell: Neural Image Caption Generation with Visual Attention". In: *Proceedings of the 32nd International Conference on Machine Learning*. 2015, pp. 2048–2057.

248. Jianyi Yang et al. "Improved Protein Structure Prediction Using Predicted Interresidue Orientations". In: *Proc. of the National Academy of Sciences* 117 (2020), pp. 1496–1503.

249. Ekim Yurtsever et al. "A Survey of Autonomous Driving: Common Practices and Emerging Technologies". In: *IEEE Access* 8 (2020), pp. 58443–58469.

250. Neil Zeghidour et al. *Fully Convolutional Speech Recognition*. 2018. arXiv: 1812.06864 [cs.CL].

251. Matthew D. Zeiler and Rob Fergus. "Visualizing and Understanding Convolutional Networks". In: *ECCV*. 2014.

252. Heiga Zen, Andrew W. Senior, and Mike Schuster. "Statistical Parametric Speech Synthesis Using Deep Neural Networks". In: *IEEE International Conference on Acoustics, Speech and Signal Processing* (2013), pp. 7962–7966.

253. Liwen Zhang, Gregory Naitzat, and Lek-Heng Lim. "Tropical Geometry of Deep Neural Networks". In: *Proceedings of the 35th International Conference on Machine Learning*. Vol. 80. Proceedings of Machine Learning Research. 2018, pp. 5824–5832.

254. Xiangyu Zhang et al. "ShuffleNet: An Extremely Efficient Convolutional Neural Network for Mobile Devices". In: *IEEE/CVF Conference on Computer Vision and Pattern Recognition* (2018), pp. 6848–6856.

255. Yong Zhao et al. "Domain and Speaker Adaptation for Cortana Speech Recognition". In: *IEEE International Conference on Acoustics, Speech and Signal Processing (ICASSP)* (2018), pp. 5984–5988.

256. Alex Zhavoronkov et al. "Deep learning enables rapid identification of potent DDR1 kinase inhibitors". In: *Nature Biotechnology* 37 (2019), pp. 1038–1040.

257. Jun-Yan Zhu et al. "Unpaired Image-to-Image Translation Using Cycle-Consistent Adversarial Networks". In: *IEEE International Conference on Computer Vision (ICCV)* (2017), pp. 2242–2251.

258. Barret Zoph et al. "Learning Transferable Architectures for Scalable Image Recognition". In: *IEEE/CVF Conference on Computer Vision and Pattern Recognition* (2018), pp. 8697–8710.

Index

Printed in the United States
by Baker & Taylor Publisher Services